# ADVANCED AUTOMATION TECHNIQUES IN ADAPTIVE MATERIAL PROCESSING

# ADVANCED AUTOMATION TECHNIQUES IN ADAPTIVE MATERIAL PROCESSING

Editors

## XiaoQi Chen
Gintic Institute of Manufacturing Technology

## Rajagopalan Devanathan
Nanyang Technological University

## Aik Meng Fong
Gintic Institute of Manufacturing Technology

**World Scientific**
*New Jersey • London • Singapore • Hong Kong*

*Published by*

World Scientific Publishing Co. Pte. Ltd.

P O Box 128, Farrer Road, Singapore 912805

*USA office:* Suite 1B, 1060 Main Street, River Edge, NJ 07661

*UK office:* 57 Shelton Street, Covent Garden, London WC2H 9HE

**British Library Cataloguing-in-Publication Data**
A catalogue record for this book is available from the British Library.

**ADVANCED AUTOMATION TECHNIQUES IN ADAPTIVE MATERIAL PROCESSING**

ISBN 981-02-4902-0

Printed in Singapore by World Scientific Printers (S) Pte Ltd

# PREFACE

Researchers and engineers have long been inspired to innovate automation solutions to a whole range of manufacturing processes such as components assembly, marking, cutting, forming, joining, grinding, polishing and chamfering. The past few decades have witnessed widespread applications of CNC technology for machining operations and robotic technology for assembly operations. Robots have also gained a solid ground in welding automation, notably spot welding for the automobile industry and arc welding for the manufacture of steel products. However, manufacturing processes that require operator's instincts and skills have posed a great challenge to researchers in the areas of robotics, automation and control. Despite many research works that have attempted to address some fundamental issues in automating material processing operations, automation practitioners find it difficult to implement integrated solutions to these applications.

The focus of this book is on how to apply and adapt advanced automation techniques and to implement integrated mechatronic automation systems. The book is intended to put in the hands of manufacturing engineers, researchers, and graduate students alike, our own as well as related recent findings on the applications of modern automation techniques. These techniques include sensors, signal processing, robotics and intelligent control for the automation of some of the demanding manufacturing processes that were traditionally handled within the mechanical and material engineering disciplines.

One of the unique features of the book is that the concepts and techniques are developed from real-life material processing applications. Specifically, the book includes the latest research results achieved through applied research and development projects over the past years in the Gintic Institute of Manufacturing Technology. The research works featured in the book are driven by the industrial needs. They combine theoretical research with practical considerations. The techniques developed have been implemented in industries.

The book begins with an overview of material processing automation. It highlights the needs of multi-facet mechatronic solutions for adaptive material processing, surveys various sensors, and explains the progression from conventional numerical control to sensor-based machine control and intelligent control.

Chapter 2 discusses process development and approach for 3D profile grinding and polishing. Simplified force control models are presented, followed by conventional model-based robotic machining. Having analysed part variations and process dynamics, the chapter details the system concept – the architecture of adaptive robotic system for 3D blending. It further presents the results of experimental process optimisation that are encapsulated in the process knowledge base.

Chapter 3 concerns the implementation of the concept presented in Chapter 2. It discusses finishing robots, control interface, in-situ profile measurement techniques, the template-based optimal profile fitting (OPF), and adaptive robot path planner (ARP). The OPF algorithm finds the best fitting of design data to the measurement points. It is fast in convergence, and robust. The ARP engine generates the robot path points, and associates them with optimal process parameters. Testing results are presented to benchmark the implemented system against the stringent requirements.

Chapter 4 focuses on acoustic emission sensing for machining monitoring and control. Following a brief overview on sensors in machining process monitoring, it explains the acoustic emission sensing mechanisms and experimental set-up. It further discusses signal processing techniques such as time domain, frequency domain, and time and frequency domain analysis methods. In particular, the wavelet analysis method has been applied for tool condition monitoring.

Chapter 5 looks into techniques of automatic weld seam tracking, specifically for advanced welding operations. Through-arc-sensing technique, robotic system integration, PID control and test results are presented. When weld preparation geometric data are required for vision-based seam tracking and welding parameter control, optical vision sensors provide a practical solution.

To complement Chapter 5, Chapter 6 addresses weld pool geometry sensing. It briefly surveys a number of weld pool sensing techniques including weld pool oscillation, ultrasound, laser array, infrared sensing, and shape depression sensing. In the development work, the laser strobe vision was employed to sense the dynamic change of the weld pool topology. A fuzzy logic (FL) controller takes the feedback information of the pool geometry and automatically adjusts the welding parameters to achieve the desired weld formation.

Chapter 7 presents the implementation of an integrated robotic gas tungsten arc welding (GTAW) system. It discusses the system architecture and sub-systems, manipulator configuration, kinematics analysis and simulation, process control, open-architecture CNC controller, and remote control techniques.

Finally, Chapter 8 extends the discussion to emerging and existing laser-based material processing applications. Laser equipment, typical applications and some automation areas are introduced. The chapter analyses optical and acoustic signals emitted from the plasma, and presents the sensor design. It further discusses signal processing though Fast Fourier Transfer (FFT) and wavelet analysis. Real-time monitoring of laser welding using an artificial neural network (ANN) is presented.

The collection of the research results in this book could not be possible without strong support from the management of Gintic Institute of Manufacturing Technology which has allocated valuable resources towards the projects, from which most of the information and results reported in the book have originated. The authors would like to thank the National University of Singapore (NUS) and the Nanyang Technological University (NTU) for collaborating with Gintic in some of the research projects. They acknowledge the funding agencies, the Economic Development Board (EDB) and the Agency for Science, Technology and Applied Research (ASTAR), for providing research grants to the research activities. The authors would also like to express their sincere appreciation to their industrial partners, Singapore Technology Kinetics (STK) and Turbine Overhaul Services Private Limited (TOS), for their entrusting Gintic with challenging projects, their co-operation and valuable inputs. Finally special thanks are due to colleagues and friends for their technical and administrative support in the research projects, and kind assistance and valuable comments in reviewing the manuscript.

# CONTENTS

# CHAPTER 1

# OVERVIEW OF MATERIAL PROCESSING AUTOMATION

Aik Meng Fong, XiaoQi Chen, Huaizhong Li

*Gintic Institute of Manufacturing Technology,*
*71 Nanyang Drive, Singapore 638075*

## 1. Constrained and Non-Constrained Material Processing

Material processing refers to those manufacturing processes that are employed to change the geometry and properties of a workpiece to achieve the desired final products or components. Changes in certain properties may be desired for specific functionalities. For example, cutting tools may be hardened to prolong their lives through heat treatment. However, we focus on material processing automation techniques for geometrical changes of a workpiece in terms of dimensions, shapes, and surface finish.

In general, a product can be made of non-metallic, semiconductor, or metallic materials. Our discussion is confined to material removal and joining of metallic parts. It is worth noting that new engineering materials such as superalloys have been increasingly introduced to meet the harsh working environment and demanding design requirements.

For example, Titanium has been pervasive in fabrication of lightweight and corrosion-resistant structures. Superalloys such as Inconel have been widely used for aircraft engine components thanks to their superior high-temperature strength. Processing of these materials is more complex and difficult than conventional engineering materials, hence requires advanced automation techniques.

Material removal may involve machining, cutting, grinding, polishing, chamfering, etc. Typically, Computer Numerical Control (CNC) machines have been dominant in these application domains. But increasingly robotic machining has advanced into material removal applications that require a high degree of dexterous manipulation and intelligent sensing and control. As far as material joining is concerned, bonding or welding is formed between two workpieces that are of same material or dissimilar materials through thermal fusion. Industrial robots or customised robots are predominantly employed for such tasks.

Depending on the nature of material processing, different types of energy sources can be employed. Table 1 lists some energy sources and their typical applications for materials processing.

Table 1 Some energy sources and material processing applications.

| Energy sources | Typical applications |
|---|---|
| Mechanical: hard cutters, grinding tools, polishing belts, etc. | Cutting, grinding, polishing, chamfering |
| Water jet | Cutting, trimming, cleaning |
| Gas flame | Cutting, brazing |
| Convection heating | Diffusion, brazing |
| Electric arc | Welding, metal fusion, deposition |
| Plasma | Welding, coating |
| Laser | Cutting, marking, trimming, welding |

## 2.      Multi-Facet Mechatronic Automation

Broadly speaking, material processing can be categorised as constrained and non-constrained. The latter involves minimal or no force interactions between the tool and the material being processed. Examples are welding, marking, dispensing, spraying and painting. The former typically involves contact forces (cutting forces) between the tool and the material. In constrained material processing such as milling and turning, the amount of material removal is largely controlled by tool position (path points), though tool wear does affect the material removal. On the other hand, for finishing operations such as grinding and polishing with soft tools, the material removal is primarily controlled by the contact force that is in turn determined by the tool characteristics and path points. Figure 1 summarises typical material processing applications, and automation components required.

In a material processing application, there may exist static and dynamic errors. Static errors may include machine errors, tooling inaccuracies and part tolerances. They can be overcome by datum detection and machine calibration. Typically tooling errors may cause the actual workspace to deviate from the ideal workspace that is modelled in the computer environment off line. Consequently the off-line program cannot work in the actual environment. However, it can be transformed into the actual workspace if the datum is detected through some sensor means. Machine errors due to fabrication process or wear, such as non-parallelism and backlash, can be compensated through calibration.

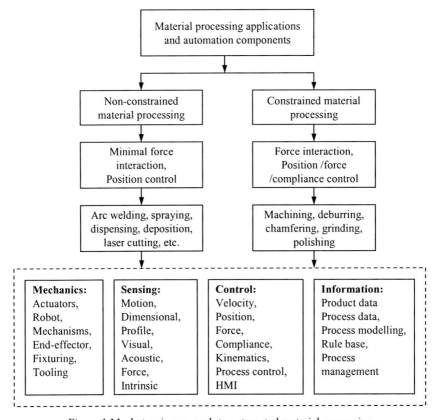

Figure 1 Mechatronic approach to automated material processing

In many demanding material processing applications, dynamic errors exist. For example, thermal distortion occurs during the welding operation.

It may severely affect the weld bead formation and quality. In machining, grinding and polishing difficult-to-machine materials, tool wear is a dynamic factor that affects the final finish. In addition, tool-workpiece vibration, especially chattering, is also a common detrimental phenomenon and a major obstruction in achieving higher productivity and better accuracy. These processes, requiring on-line compensation for dynamic errors to achieve desired results, are termed as adaptive material processing. The dynamic errors cannot be solved by conventional robot control or Computer Numerical Control (CNC) technology. Instead they call for mechatronic solutions that are typified by synergistic integration of customised mechanics, sensors, modelling and control, and product and process information, as illustrated in Figure 1.

## 3.       Sensors for Material Processing

### 3.1     *Measurands in Material Processing*

A sensor converts energy from one form to another. It is understood that there exist ten forms of energy: atomic, electrical, gravitation, magnetic, mass, mechanical, molecular, nuclear, radiant and thermal. In the context of material processing discussed in this book, the following four forms of energy are relevant.

- Electrical energy. It is related to electrical field, current, voltage and power.
- Mechanical energy. It pertains to motion (velocity, acceleration), displacement, force, torque, etc.
- Radiant energy. It is related to infrared, visible light, ultraviolet and X-rays, etc.
- Thermal energy. It is related to the kinetic energy of atoms and molecules.

Corresponding to the above forms of energy, there are four types of signals: electrical signal, mechanical signal, radiant signal and thermal signal, which can be measured to gain insight information about a process being monitored and controlled.

The input quantities, properties or conditions that are detected or measured by sensors are called measurands. For the interests of material processing, the common measurands are described as follows.

## Position

The measurand is detected for part locations or geometric information. Contact probe sensors have been widely used in many industrial applications for a long time, especially owing to their simple design and easy handling and maintenance. For decades, contact sensors have been in use to determine the dimensional changes of workpiece during manufacturing. In welding, contact sensors permit the detection of the welding start/end points and tracking weld seam, with comparatively low technical expenditure. Another special contact sensor is an electrode or wire contact sensor that was developed for arc welding robots. The main limitation of the use of contact probe sensors is the wear of the probe itself. The measuring speed and precision are also limited. Increasingly non-contact optical displacement transducers have been used for factory automation. By incorporating extra degrees of motion, 3D geometrical information can be readily obtained.

## Vibration

Vibration can be measured by an accelerometer. In early approaches, strain gauge accelerometers are used which have a seismic mass suspended from thin flexure plates, and their frequency ranges are quite low. Capacitive accelerometers also have distinct advantages for low frequency, low level sensing. The piezoelectric accelerometers are more prevailing currently. They have the advantage of robustness, simplicity and small size. They are best suited for measurement of high frequency, high-level acceleration, as found in impact testing.

## Acoustic Emission

Acoustic emission (AE) is a class of phenomena whereby transient elastic waves are generated by the rapid release of energy from localised sources within a material. In cutting, the most important sources of AE are friction at the rake face, friction between workpiece and tool, plastic deformation in the shear zone, chip breakage, contact of chip with either workpiece or cutting edge, and crack formation. AE is normally measured using a conventional piezoelectric AE transducer. Dynamic microphones can also be used to pick up sound waves with frequency from a few Hz to a hundred kHz. Two types of AE sensors have to be distinguished, wide-band sensors and resonance systems. For machining operations, AE sensor based monitoring systems are commercially available, which can be used for the detection of tool breakage, tool wear, chatter occurrence, etc. They can also be used to detect laser plasma reaction on a workpiece.

## Force

Monitoring of cutting forces is highly desirable since the process behaviour of the material removal is reflected by the change in the cutting forces. These quantities can be used to monitor tool condition (tool wear, tool breakage/chipping) and cutting condition (chatter vibration). Strain gauges, piezoelectric quartz force transducers, and dynamometers are available for force measurement.

## Torque

For milling and drilling processes, the measurement of torque is sometimes needed. The most effective and simple method for the measurement of torque on static components is to use a torque arm and a force transducer. Torque transducers, generally strain gauge or piezoelectric based, are available for mounting on machining spindles and transmit their output to the instrumentation either by slip rings or by radio signals. However, the large engineering effort and the additional space required are limiting factors. Other solutions include the non-contact magnetostrictive sensor that is regarded as the most promising low-cost version for industrial use as it does not need any major changes to the machine set-up.

## Power and Current

The measurement of the spindle or axial drive power, which can provide an indication of tool wear or breakage, can be regarded as technically simple. From the measurements of current, line voltage, and phase shift, motor power can easily be calculated. It is even possible to gain information about the actual power demand of the drives from the machine tool control without additional sensors. However, the sensitivity of this measuring quantity is limited because the power required for cutting is only a portion of the total power consumption. Furthermore, the slow response speed is a severe limitation in many machining applications.

Current is very useful in detecting welding torch location once the torch is oscillated across a groove. It is because the arc current is inversely proportional to the distance between the contact tip of the torch and the workpiece.

## *Temperature*

A thermocouple can be integrated in the tool or workpiece for point temperature measurement. If several thermocouples or different measurement positions are applied, a temperature distribution in an area of interest can be determined. This distribution can also be obtained by infrared imaging, video-thermography, or total radiation pyrometry. However, the industrial use of these sensors as a means of process monitoring is rather limited.

## *Light Intensity*

In automatic welding, laser processing, and micro-machining, visual-based sensors are widely used to identify the workpiece-related geometric values and to monitor the process performance. CCD camera based imaging systems are used as devices for seam tracking, groove shape detection, and detecting weld start/end points. Some optical components and laser diodes are also needed in these devices. In welding, keyhole diameter and weld pool dynamics can be monitored by using CCD cameras.

## *Pressure/Flow*

Sometimes sensors for coolant supply monitoring are needed. Coolant pressure and flow rate, measured with a simple flowmeter in the coolant supply tube before the nozzle, are now often part of a process parameter description.

## 3.2    *Types of Sensors*

A variety of sensors can be utilised to monitor the measurands of interest. They can be classified according to their sensing principles.

- Resistively coupled transducers - potentiometer, strain gauge, thermometer, dynamometer;
- Variable inductance transducers - Linear Variable Differential Transducer (LVDT), Rotary Variable Differential Transducer (RVDT), resolver, etc;

- Variable capacitance transducers - capacitive gauge, angular gauge;
- Permanent magnetic transducers - AC/DC tachometer;
- Eddy-current transducers - proximity sensors;
- Piezoelectric transducers - accelerometer, acoustic emission, microphone, velocity sensor, dynamometer, etc.;
- Optical transducers - fiberoptic sensor, interferometer;
- Ultrosonic transducers - imaging scanner, displacement sensor;
- Piezoresistive transducers - stress and force measurement;
- Photovaltaic transducers - detecting illumination;
- Megnetoresistive transducers - measure magnetic field measurement;
- Thermocouples - temperature measurement;
- Charge-coupled device - imaging and displacement measurement.

Table 2 lists some sensor manufacturers and their products and services.

## 3.3   *Microsensors and Soft Sensors*

The rapid growth of silicon microsensors based on micro electromechanical systems (MEMS) technology is an important factor in the implementation of sensors in manufacturing. This technology allows the integration of traditional and novel sensing technologies onto miniaturised platforms, providing in hardware the reality of multi-sensor systems. In addition, these sensors are easily integrated with the electronics for signal processing and data handling on the same chip, hence making the sophisticated signal analysis including feature extraction and intelligent processing more straightforward and inexpensive. The combination of microelectronics and microsensors in instrumentation design, which has greatly increased device functionality, has given rise to a generation of devices referred to as smart instrumentation.

The demand for improving the performance of the monitoring system is increasing. Sensor fusion is a powerful tool for making the monitoring system more reliable and flexible. In a replicated sensor system, the integration of similar types of sensors can contribute to the improvement of the reliability and robustness of the monitoring system, whereas in a disparate sensor system, the integration of different types of sensors can make the monitoring system more flexible. The advancement of fast data processing technologies makes a monitoring system more practical in the manufacturing environment, while soft computing techniques such as fuzzy logic, artificial neural networks and genetic algorithms can contribute to the intelligence of the monitoring system.

Table 2 Products and services of some sensor manufacturers.

| Manufacturers | Products & Services |
|---|---|
| Dunegan Engineering Consultants Inc. | Supply acoustic emission equipment, academic and industrial news and product information. The main products are AESMART series AE instrument and various AE sensors. |
| Montronix Inc | Provide complete monitoring systems as well as various kinds of sensors: power sensor, torque sensor, force sensor and broadband vibration sensor et al |
| Prometec GmbH | Provide modular tool and machine monitoring system for turning, drilling, milling and grinding. Supply force sensors, axial displacement sensor, AE and vibration sensor |
| Kistler Instrument | Pressure sensors, force sensors, acceleration sensors, AE sensors. Kistler dynamometers are very popular for force measurement in machining process. Also provide rotating dynamometer and digital telemetry. |
| Bently Nevada Corporation | Provide a comprehensive scope of products encompassing vibration monitoring instrumentation, lubricant condition monitoring, and fundamental rotor dynamic research. |
| Brankamp | Supply machine and tool protection for all kinds of production machines. Provide process monitoring systems, special monitoring programs and sensors. |
| Keyence | Provide a variety of sensors for industrial applications. Photoelectric sensors, displacement (CCD laser, confocal, inductive, ultrasonic), area sensors, proximity, through-beam measurement, pressure, and vision sensors. |

Another emerging sensing technique is soft sensor. Soft sensors provide a way of indirectly measuring values when it is not possible to take a direct measurement. In fact, soft sensors can be neural network based programs that calculate difficult-to-measure values based on other, easier-to-measure values. Today, many software vendors of advanced control provide soft sensor technology in the form of configurable packages that are relatively easy to use.

## 4.      Intelligent Control Techniques

### 4.1    *Conventional Computer Numerical Control*

Controllers are designed to automatically control process variables such as motion, temperature, or pressure. A controller accomplishes this by changing a process input so that a process output agrees with a desired result: the setpoint. This is typically achieved through a combination of feedback and feedforward control.

An automated machining system comprises CNC controller, machine tool, and machining process. The motion of the machine tool is achieved by coordinated control of servomotors. The CNC controller sends out commands for the movement and the result is continuously monitored and fed back through various devices to ensure that the relative motion between the tool and the workpiece follows the predetermined command motion. Generally there are two types of feedback loops: the inner velocity feedback and outer position feedback loops, as shown in Figure 2. The feedback is obtained through motion sensors like encoders, tachometers, or linear scales.

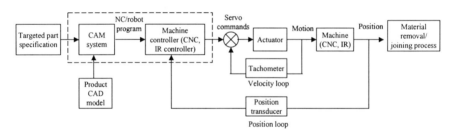

Figure 2 Basic control loops in a CNC machining system.

The most common approach in conventional CNC is the proportional-integral-derivative (PID) control. An ideal PID controller has terms that compensate for the instantaneous deviation from target, long-term persistent deviation from target, as well as the rate of change of the deviation from target. Hence such a controller has three tuning parameters: the proportional gain, integral time and derivative time. Conventional approaches to tuning PID controllers are empirical in nature. Model-based tuning approaches have since attracted considerable research interests to meet increasingly higher performance demands.

Commercial PID controllers have evolved into an instrument of great versatility over decades. As an improvement to the basic PID control, further control strategies, such as ratio control, lead-lag control, cascade control, dead-time compensation, and feedforward control, have been developed. Efforts in the last few decades have focused on non-linear control. It is interesting to note the increasing dependence of regulatory systems on a process model as they have evolved.

PID-based control systems are most commonly used in single-input-single-output (SISO) processes. However, not all industrial processes can be controlled with PID loops. Most industrial processes are multivariable in nature. That is, there are usually many process variables to be controlled using a set of actuators. Furthermore, there exist significant interactions between the process variables and control actuators. In these situations, single-loop regulatory controllers will typically produce poor performance. Multivariable, non-linear, and random processes all require more advanced control techniques.

In addition, the traditional control techniques for CNC machine tool might not be adequate in the modern highly automated manufacturing environment due to the following reasons:

- The primary function of CNC is to automatically execute a sequence of multi-axis motions according to the part geometry. However, safe, optimal, and accurate machining processes are planned by manufacturing engineers based on their experience and understanding of the process. An intelligent CAM system is needed which is capable of predicting optimal tooling and cutting parameters to maximise the cutting efficiency based on process models.
- There are only two basic feedback devices in a conventional normal motion control system: position and velocity feedback transducers. It is rather difficult to predict vibration, tool wear and breakage, thermal deformation of the machine tools, and similar processes based on events using off-line theoretical models. Hence, the state-of-the-art CNC technology guarantees correctly controlled feeds, speeds, and positions, but not end results.
- There is a need for newer control systems capable of adjusting the cutting parameters automatically to achieve better machining quality and higher productivity. These cutting parameters may include feed rate, depth of cut, spindle speed, etc. The adjustment is based on the

process model and the cutting condition signals acquired on line by a set of sensors.

## 4.2    *Sensor Based Machine Tool Control*

A solution to intelligent manufacturing is sensor-based machine tool control. Besides the normal feedback loop of velocity and position, the machining process data can be used as the third feedback loop for process control, as shown in Figure 3. The machine tool is instrumented with in-process sensors to measure force, vibration, displacement, temperature, acoustic emission, as well as vision signals on line. These measured signals are processed by real-time monitoring and control algorithms to detect overload, chatter, vibration, tool wear, and tool breakage. Through adaptive process control, corrective actions are taken by the CNC accordingly, such as the manipulation of spindle speed, feed, and tool offsets, and compensation for geometric variants and process dynamics.

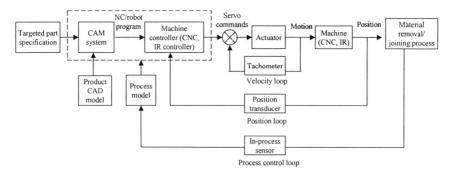

Figure 3 Control loops in a sensor-based intelligent machining system.

Modern communication networks enable economical and fast data communication from sensors to the controller. With the recent advances in computer and sensor technology, combining sensors with an open architecture PC-based CNC machine tool controller becomes more viable. Simple sensors, such as tool breakage monitors, are already part of a machine tool control system. 3D vision systems and image processing techniques are being studied to perform in-cycle part measurement as well as tool flank wear monitoring. For improved machining performance, sensor fusion techniques have been applied to intelligent manufacturing studies, and neural network and expert system techniques have been used as the methodologies to interpret the data from multiple sensors.

In an intelligent machining system, the sensors should have a sufficient bandwidth and be practical for installation on machine tools. To realise real-time sensing and monitoring, efficient and fast signal processing is essential. Effective process models also need to be established. Intelligent manufacturing technologies developed in laboratories are rapidly progressing towards industrial adoptions. However, there still exist some hindering factors such as lack of robust sensor hardware, reliable process models, as well as standardisation. Further work is needed in areas such as multi-sensor fusion and knowledge-based recognition to deal with the divergent factors in the machining process.

## 4.3　*Open Architecture and Distributed Control*

Most commercial CNC systems are designed as closed architectures and supported by proprietary technologies. These CNC systems cannot be easily modified to accommodate specialised control needs. It is virtually impossible to incorporate different types of control routines, such as substituting a standard PID control with an adaptive control. It is also very difficult to incorporate different types of sensor based control schemes into these CNC systems.

To achieve intelligent adaptive material processing, the architecture of CNC must be organised in such a way that it allows real-time manipulation of the machine tool's operating conditions. It demands open-architecture CNC systems that facilitate modular design and integration of user-developed sensor-fused real-time application programs into the machine tool controller. Open architecture control systems allow the most suitable components from different vendors to be integrated harmoniously and seamlessly. Furthermore, incremental enhancement of machines by employing new sensors and control methods at minimal cost becomes possible.

The level of "openness" with the "open" controllers advocated by different vendors varies. A fully open system should have the ability to modify the motion control software. Modern open architecture control systems employ digital signal processor (DSP) based multi-axis motion controllers. Many of these controllers are VME or PC based. PC-based controllers have the advantages of low cost, ease of maintenance and upgrading, large data storage, file management capabilities, networking, programmable graphical user interface (GUI), as well as compatibility with a wide range of software based control packages. Hence, PC-based controllers becomes more appealing to automation researchers and engineers.

PC-based open machine tool controllers come in two types. One type employs a PC connected to a standard proprietary CNC system, while the other consists of a PC chassis with a proprietary CNC board installed. GE Fanuc Automation provides the interface between the PC and CNC. Allen-Bradley's PC-based CNC offers open interface for servo, part programming, communication, HMI, logic, and I/O control. The open hardware architecture takes advantage of the economics of scale of the PC industry.

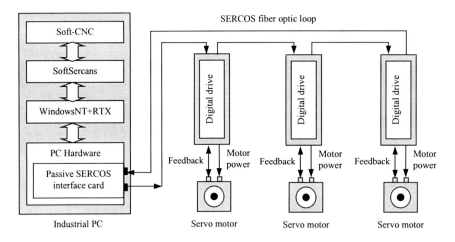

Figure 4 Block diagram of a SERCOS-based soft CNC system.

Standard device networks, such as SERCOS and DeviceNet, integrate actuators, sensors and controllers in a simpler way. They shift the machine interface from analogue to digital; avoiding tedious wiring, and enhancing the system reliability and robustness. Such an emerging paradigm reduces costs of wiring, installation, servo tuning, and upgrading. Recently open software CNC systems have emerged on the horizon. For example, MDSI has developed OpenCNC software that requires no proprietary hardware or motion control card, but a real-time operating system is needed. With an unbundled software solution, customers can pick their own hardware to obtain the best cost. Figure 4 shows a SERCOS-based soft CNC system.

## 4.4    *Intelligent Control and Computing Techniques*

One of the first and very popular advanced control techniques commercialised was model-predictive control which computes process interactions, overcomes disturbances, and predicts future performance using a process model. However, processes that are poorly understood and those that demonstrate variable behaviour are particularly difficult to model, and alternative methods have to be used.

In recent years, intelligent control techniques have been extensively studied. Many of these techniques derive from the rapidly expanding field of artificial intelligence, such as fuzzy logic, neural networks, genetic algorithms, case-based reasoning, term subsumption systems, intelligent agents, and so forth. These technologies empower manufacturing systems with sophisticated sensor integration, high level of machine intelligence, and extendable control capabilities and functionalities.

Fuzzy logic (FL) attempts to build a control system that works by classifying and dealing with data in a similar way to human beings. FL puts the experience of engineers and operators directly into a control solution. The challenge is to find an adequate way to formulate and incorporate operator knowledge. FL categorises data by probabilities, and works with linguistic rules (uncertainty) and partial truth. It allows supervisory control strategies to be defined using the same rules that humans would. Even complex applications can be solved with only a few rules because FL lets designers use "rules-in-general." Up to date, FL has the most industrial applications of all the artificial intelligence technologies.

Artificial neural networks (ANNs) are mathematical algorithms inspired by studies of the human brain and nerve system. They encompass various techniques for modelling non-linear processes. ANN algorithms are rooted in statistics and they can learn by training. Control is one of the newer and possibly the highest return areas where ANNs have been applied. ANNs are particularly helpful for analysing processes that cannot be quantified by traditional linear models. In situations involving uncertain process data, ANNs infer the current value of the process variable from a collection of related variables. Using historical data, a neural network can be "trained" to "learn" the complex relationships among the measurable variables and the variable of interest. Once these relationships are known, the current value of the inferred variable can be calculated from those of the variables that are actually measured.

The most popular ANN used in control applications is Backpropagation (BP), also referred to as Multi-Layer Perceptrons or Feedforward Neural Networks. Backpropagation is a very general modelling technique that intrinsically captures the physical property of smoothness. Its ability to develop non-linear models from data has been applied in model-predictive control, inferential sensing, sensor validation, process optimisation, and product formulation.

An expert system (ES) conducts its reasoning based on rules and experience derived from experts. An expert system for machining control works by complementing the process models. It makes decisions experts' knowledge and rules encapsulated in the knowledge base. ESs have been widely applied in intelligent manufacturing, from sensor validation through machining process control, tool wear monitoring, to process optimisation.

Genetic algorithms (GA) simulate biological processes. In this paradigm, a population of candidate solutions, encoded as knowledge structures, evolve by competition and controlled variation. As the population evolves, progressively improved solutions to the problem are generated. Hence GAs can utilise information about solution fitness efficiency over a very large number of possible solutions.

## 4.5    *Human-Machine Interface*

An important aspect of modern machine control is the human-machine interface (HMI) that provides interactions between human and machine or computing device. The interface can display machine status, alarms, messages and diagnostics. It provides the operator with process feedback information, or allows the operator to intervene a running process.

Recently the worldwide HMI market has seen rapid growth due to the development of new technologies and increasing needs from industry. HMI software is shifting from stand-alone computing towards client-server architecture. Many Windows NT based products have been developed with the features of open systems, client/server models, and Internet communications, which promise faster information access, greater agility, and lower costs. Some vendors produce common classes of software packages such as APIs. It provides users with the ability to monitor and supervise a control system through a configurable platform.

In an intelligent manufacturing system, HMI will continue to serve as a centralised system for coordinating different aspects of the manufacturing process, from commands input, sensing signal acquisition and processing, machining condition monitoring, alarming, to adaptive machine tool control.

# References

1. Chen, X. Q., Gong, Z. M., and Huang, H. "Development of adaptive robotic system for 3D profile grinding / polishing", Gintic Technical Report AT/00/012/AMP, 2000.

2. Chen, X.Q., Zeng, H., and Wildermuth, D. "In-process monitoring through acoustic emission sensing", Gintic Technical Report AT/01/014/AMP, 2001.

3. Wang, J.Y., Mohanamurthy, P.H., Foong, M.K., Devanathan, R., Chen, X.Q., and Chan, S.P. "Development of a closed-loop through-the-arc sensing controller for seam tracking in gas metal arc welding", Gintic Technical Report AT/00/013/AMP, 2000.

4. Shacklock, A., Luo, H., Huang, S., and Wang, J. "Intelligent robotic GTAW system for 3D welding", Gintic Technical Report AT/01/013/AMP, 2001.

5. Sun, Z., Pan, D., & Kuo, M. "Precision welding for edge build and rapid prototyping", Gintic Technical Report PT/99/001/JT, 2000.

6. Sun, Z., Kuo, M., and Pan, D. "Twin wire gas tungsten arc cladding", Gintic Technical Report PT/99/004/JT, 2000.

7. Gong, Z.M., Huang, S., and Zeng, H. "Development of PC-based Adaptive CNC Control System", Gintic Technical Report AT/01/043/AMP, 2001.

8. Knopf, G.K., Muir, P.F., and Orban, P.E. (editors), "Sensors and controls for intelligent machining and manufacturing mechatronics," Proceedings of SPIE, Boston, Massachusetts, USA, 19-20 September 1999.

9. Liang, S.Y. "In-process sensing & signal processing for monitoring and control of machining," Gintic – Georgia Tech Joint Workshops on Manufacturing technology, Technical Workshop 4, 26-27 September 2001, Singapore.

10. Tönshoff, H.K., and Inasaki, I. "Sensors Applications Volume 1: Sensors in Manufacturing", Wiley-VCH, Verlag GmbH, 2001.

11. Zhou, Y., Orban, P., and Nikumb, S. "Sensors for intelligent machining – a research and application survey," 0-7803-2559-1/95, 1995 IEEE, pp. 1005-1010.

12. Parr, E.A. "Industrial Control Handbook", 3rd Edition, Newnes, 1998.

13. Forbs, J.F. "Process control: from PID to CIM," in Materials Processing in the Computer Age III, Edited by Vaughan R. Voller and Hani Henein, The Minerals, Metals & Materials Society, 2000, pp. 15-25.

14. Altintas, A. "Manufacturing Automation", Cambridge University Press, 2000.

15. Zuo, J., Chen, Y.P., Zhou, Z.D., Nee, A.Y.C., Wong Y.S., and Zhang Y.F., "Building open CNC systems with software IC chips based on software reuse," International Journal of Advanced Manufacturing Technology, 2000, pp. 16:643-648.

16. Ulmer, B.C.Jr., and Kurfess, T.R. "Integration of an open architecture controller with a diamond turning machine," Mechatronics (1999), pp. 349-361.

17. Bartos, F. J. "Artificial intelligence: smart thinking for complex control," Control Engineering Online, July 1997.

18. VanDoren, V.J. "Advanced control software goes beyond PID," Control Engineering Online, January 1998.

# CHAPTER 2

# PROCESS DEVELOPMENT AND APPROACH FOR 3D PROFILE GRINDING/POLISHING

XiaoQi Chen*, Zhiming Gong*, Han Huang*, Shuzhi Ge**, Libo Zhou***

*Gintic Institute of Manufacturing Technology,
71 Nanyang Drive, Singapore 638075*

**Department of Electrical & Computer Engineering, The National University of
Singapore, 10 Kent Ridge Crescent, Singapore 119260*

*** Department of System Engineering, Ibaraki University, Japan*

## 1.    Introduction

Industrial robots are gaining widespread applications in manufacturing process automation. Their applications can be classified into two broad categories, namely non-constrained and constrained manipulation. The former does not involve force interaction or control between the end-effector and the environment that the robot acts on. Examples of non-constrained processes are inspection, laser cutting, welding, plasma spraying and many assembly tasks. Typically, the position control with or without external sensors is sufficient to accomplish these tasks. On the other hand, constrained robotic tasks such as machining, deburring, chamfering, grinding, and polishing involve force interaction between the tool and the workpiece to be processed. In addition to position control, the contact force and process parameters must be controlled to achieve the desired output.

For the past decade the theory of force control for constrained robotic applications has been extensively researched [1]. By and large, there are three approaches to force control, and these are impedance [2], hybrid position/force control [3] and constrained motion control [4]. Instead of tracking the motion and force trajectory, the impedance control regulates the dynamic behaviour between the motion of the manipulator and the force exerted on the environment. In [5], the impedance control of robot manipulators using adaptive neural network is proposed. Without an explicit force error loop, the desired dynamic behaviour is specified to obtain a proper force response.

Hybrid position/force control combines force and torque information with positional data to simultaneously satisfy position and force trajectory constraints that are specified in a task-related coordinate system. In [6], an adaptive controller for force control with an unknown system and environmental parameters is examined. Force control, impedance control, and impedance control combined with a desired force control are treated using model-reference adaptive control (MRAC). A recent work [7] has examined the stability of the most basic hybrid control, which requires no robot dynamic model. Many other researchers have studied various learning methods for hybrid force/position control [8-12]. Attempts have been made to apply constrained robotic control for deburring and chamfering [13-18]. A research work on polishing sculptured surface using a 6-axis robot is reported [19]. Generally speaking, constrained robotic applications are largely confined to laboratory explorations, particularly for an unknown environment. Only a handful of commercial systems such as Yamaha finishing robots are available for 3D profile polishing. Furthermore they are limited to processing new parts in simple operations, which rely on teaching and play-back or CAD-driven off-line programming.

One of the hurdles in automating constrained robotic tasks is that it is difficult, indeed very often impossible, to derive an analytical model to describe the process to be controlled. Successful execution of the empirical process relies heavily on human knowledge. The problem escalates when part and process variables must be considered, such as part distortions (typical in aerospace component overhaul), severe tool wear (most pronounced in processing superalloys such as Inconel), and process optimisation to meet stringent standards required by the industry. All these factors must be vigorously studied before proposing a feasible approach to such an intriguing problem.

This chapter discusses the perspective and approach of 3D profile grinding and polishing in general, and blending of overhaul jet engine components specifically. The JT9D first stage turbine vane is selected as a case study. Section 2 discusses the surface finishing processes including manual blending, CNC milling and wheel grinding. Consideration of poor machinability of the material leads to the determination of a suitable process to automate the profile blending operation. Section 3 establishes the model of the contact force between the tool and the workpiece, which is crucial to desired material removal and surface finish. Section 4 generalises model-based robotic machining systems, highlighting their advantages and deficiencies. Section 5 details part variations and process dynamics that are predominant in robotic blending. Section 6 proposes a knowledge-based adaptive robotic system by integrating various intelligent software modules, such as Knowledge-Based Process Control (KBPC) and Data-Driven Supervisory Control (DDSC). Section 7 presents results of tool path optimisation and tool wear compensation, which can be incorporated into the knowledge-based process control to address process dynamics. Finally, the chapter is concluded with some remarks on the proposed approach and concept.

## 2.     Profile Grinding and Polishing of Superalloys

### 2.1     *Superalloy Components and Manual Blending*

Superalloys are widely used, in the aerospace industry for example, to meet the following demanding engineering requirements:

- High strength-to-weight ratio
- High fatigue resistance
- High corrosion resistance
- Superior high-temperature strength

Jet engine turbine vanes and blades are often made of Inconel materials. However, these materials have poor machinability, long recognised by manufacturers. Figure 1 shows the schematics of a high-pressure turbine (HPT) vane.

The vane consists of an airfoil having concave and convex surfaces, an inner buttress and an outer buttress. After operating in a high-temperature and high-pressure environment, vanes incur severe distortions as large as 2 mm in reference to the buttress. On the airfoil surface there are hundreds of cooling holes. After a number of operational cycles, defects such as fully or partially blocked cooling holes, micro cracks and corrosions begin to occur. Because of the high cost of the components, it is common practice to repair these parts instead of scrapping them. The repairing process starts with cleaning and covering the defective areas with the braze material. The purpose of brazing is to fill up the defects, but unavoidably, the brazed areas will be higher than the original surface. Figure 2 shows a cross section of the airfoil brazed with a repair material.

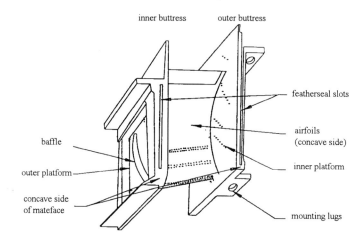

Figure 1 Schematics of a HPT Vane.

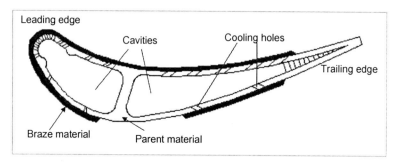

Figure 2 Turbine airfoils repaired with braze material.

Table 1 summarises the overhaul conditions of typical jet engine high-pressure turbine vanes. The wall thickness of the airfoil ranges from 0.8 mm in the trailing edge to 2 mm in the leading edge. The braze material, similar to the parent material in composition, is laid down on the airfoils manually, and its thickness is about 1 mm. The compositions of the braze material are shown in Table 2. The three major chemical elements of both parent and braze materials are cobalt (Co), chromium (Cr) and nickel (Ni).

Table 1 Overhaul conditions of jet engine turbine vanes.

| Items | Conditions |
|---|---|
| Airfoil material | Inconel |
| Airfoil curvature | 3D, concave and convex |
| Braze material | Similar to parent material |
| Hardness | 20 to 30 HRC |
| Machinability | Poor |
| Part dimension | $150 \times 140 \times 80$ mm (max.) |
| Part weight | 0.68 kg (max) |
| Wall thickness | 0.8 (trailing edge) to 2 mm (leading edge) |
| Part distortion | Up to 2.0 mm |
| Braze thickness | 0.5 – 1.5 mm |
| Braze pattern | Defined sets |
| Braze coverage | About 80% of airfoil surface |
| Brazing operation | Manually done with dispenser |

Table 2 Compositions and properties of braze material.

| Chemical elements | Atomic Number | Atomic Weight | Density $(g/cm^3)$ | Weight Percentage |
|---|---|---|---|---|
| Cobalt | 27 | 58.93 | 8.90 | 45.6 |
| Chromium | 24 | 51.996 | 7.19 | 23.75 |
| Nickel | 28 | 58.693 | 8.902 | 25 |
| Tungsten | 74 | 183.84 | 19.30 | 3.5 |
| Tantalum | 73 | 180.95 | 16.654 | 1.75 |
| Titanium | 22 | 47.90 | 4.54 | 0.1 |
| Carbon | 6 | 12.01 | 3.513 | 0.3 |
| Baron | 5 | 10.811 | 2.34 | 1.48 |

Blending can be defined as the material removal process to achieve the desired finish profile and surface finish roughness. The process is often employed to remove excessive material on surfaces of new jet engine parts or overhaul turbine airfoils. Within the blending process, we further define rough grinding as the process step to remove the excessive material with the profile generation as the primary aim. The aim of fine polishing is to achieve the desired surface roughness. In this sense, the term blending is often interchangeable with grinding and polishing.

In the aircraft overhaul industry, the blending process is intended to remove excessive braze material for the brazed area to be flush with the original surface within a tight tolerance. Current manual blending (also called belt polishing) uses belt machine to remove the braze, within tolerated undercuts and overcuts, as illustrated in Figure 3 (a). After belt polishing, the part is polished with a flap wheel, as shown in Figure 3 (b), to achieve the final surface finish. Table 3 lists quality requirements of the blending operation. They are achieved with the operator's skills and knowledge:

- Manipulate the part correctly in relation to the tool head.
- Exert correct force and compliance between the part and the tool through wrists, and control the force interaction based on process knowledge.
- Adapt to part-to-part variations through visual observation and force feedback.

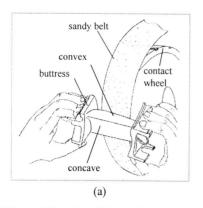 

(a)                                              (b)

Figure 3 Manual polishing of brazed airfoils with (a) belt polishing tool, (b) flap wheel.

One can imagine that a possible automation solution is to develop a machine which can mimic the operator's capabilities. In an abrasive machining process such as belt polishing, the amount of material removed not only depends on tool position, but also the contact force between the tool and workpiece. Such an automation system requires position control as well as force control so that the desirable amount of material can be removed to avoid excessive overcut or undercut. The complexity of such automation further escalates in consideration of process dynamics and part-to-part variations typified by near-net-shape new parts and overhaul parts. The first hard choice is what material removal process is most suitable for the intended automation system. Hence, the machining processes for superalloy materials must be carefully evaluated.

Table 3 Quality requirements of airfoil blending.

| Items | Specifications |
|---|---|
| Overcutting | ≤ 100 microns |
| Undercutting | ≤ 100 microns |
| Trailing edge | Absolutely no overcutting |
| Leading edge | 0 to 200 microns gap from the template. Smooth curvature. |
| Wall thickness | Greater than minimum wall thickness at specified check points |
| Surface roughness | ≤ 1.6 microns $R_a$ |
| Transition from brazed to non-brazed area | No visible transition lines |
| Blending path | No visible path overlapping marks |
| Part integrity | No burning marks |

## 2.2    *CNC Milling*

A four or five-axis CNC system with a hard cutting tool would be able to satisfy the position control required by 3D profile finishing. The material removal can depend solely on position control. Other researchers have proposed the CNC milling process for cutting superalloy materials. However, the key issue to be addressed is tool wear and tool life in processing difficult-to-machine materials such as Inconel.

To evaluate the feasibility of hard tool machining, experiments were conducted with conventional cutting tools on the sample material. The objective of the experiments was to monitor the tool conditions by measuring the cutting force, and establish the cycle time required. The following cutting conditions were applied:

- Machine tool: Hitachi Seiki (VG 45) 5 axis Machining Centre.
- WC ball end cutter (insert) – UX 30 (maximum rotation diameter 10 mm).
- Rotation speed of cutter: 1200 rpm.
- Depth of cut: 0.06 mm.
- Transverse feed: 150 mm/min.
- Length of cutting pass: 58 mm (removal rate: 1 gram/min.).

The cutting force was recorded to check the tool life. Figure 4 shows the cutting force along Z-axis as a function of the cutting pass. It is apparent that the cutting force has a large increase after 30 passes, indicating significant tool wear.

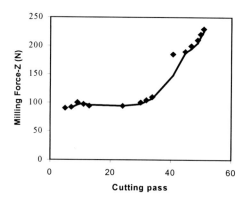

Figure 4 Milling force as a function of cutting pass.

Given that 80% of the vane surface is covered by the braze material and the braze layer is about 1.5 mm, the weight of braze material to be removed is about 70 grams. Thus the number of cutting passes required for one piece of vane is 157. This indicates that we have to change inserts 5 times for milling one vane. The effective cutting time for one vane is about 70 min. (i.e. 70 grams / 1 gram per min.). The fast tool wear would incur considerable tooling cost. The slow removal rate compares unfavourably with the manual belt polishing which takes about 10 minutes to polish away the braze material. It was therefore concluded that the CNC milling method is not suitable for the blending of turbine vane because of fast tool wear and long cycle time.

## 2.3    *Wheel Grinding*

The second process investigated was the wheel grinding process that offers an alternative solution to belt polishing. Clearly, controlling a grinding wheel with a 5-axis CNC is much easier that controlling a large polishing belt. Again, grinding wheel wear and material removal rate must be examined before introducing such a process for an automation solution.

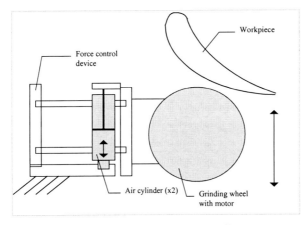

Figure 5 Experimental set-up for grinding test.

Figure 5 shows the experimental set-up of the grinding test. The grinding wheel is mounted on a pneumatic rig, which has an upward pressure and a downward pressure. The contact force between the grinding wheel and workpiece is determined by the differential air pressure of the two cylinders as follows:

$$F = (P_{up} - P_{down})A - W \tag{1}$$

where $F$ is the contact force, $P_{up}$ the upward air pressure, $P_{down}$ the downward air pressure, $A$ the surface of the piston, and $W$ the weight of the grinding motor and the grinding wheel. The experiments were conducted under the following conditions:

- $P_{up}$  : 2.0 Kgf/cm$^2$.

- $P_{down}$ : 0.4 Kgf/cm$^2$.
- Wheel rotation speed: 7546 rpm.

Three types of grinding wheels were tested:

- Norton C150D5BTM No: 06380, Size: 101.6×19.1×6.4 mm.
- Cratex 302 C (rubberised abrasives), Size: 76.2×3.2×6.4 mm.
- Cratex 304 M (rubberised abrasives), Size: 76.2.4×6.4 mm.

The test results are shown in Table 4. The three grinding wheels offer similar material removal rate between 0.13 to 0.15 grams per minute. Even if Cratex 304 M is most resistant to wear, its wear rate (0.25 g/min) is still higher than the material removal rate (0.15 g/min). For one workpiece, the total amount of material to be removed is about 45 grams. Assuming the removal rate of 0.15 g/min, the total polishing time required is 300 min.

Table 4 Material removal and tool wear for grinding wheels.

| Conditions | Workpiece | | | Grinding wheel | | |
|---|---|---|---|---|---|---|
| | Weight before (g) | Weight after (g) | Removal rate (g/min) | Weight before (g) | Weight after (g) | Wear rate (g/min) |
| Norton C150D5BTM Duration: 1.5 min. | 358.7 | 358.5 | 0.133 | 147.6 | 147. 0 | 0.40 |
| Cratex 302 C Duration: 2.0 min. | 358.3 | 358.0 | 0.15 | 42.2 | 40.6 | 0.80 |
| Cratex 304 M Duration: 2.0 min. | 358.0 | 357.7 | 0.15 | 66.2 | 65.7 | 0.25 |

The evaluation results for milling, grinding and belt polishing are summarised in Table 5. Clearly, all possesses can meet the surface finish requirements of 1.6 micron $R_a$. However, they differ greatly in terms of material removal rate, hence the cycle time. For milling and grinding processes, the removal rates are 0.15 g/min and 0.8 g/min respectively. The resultant long cycle times are unacceptable for production use when there are large amount of materials to be removed. On the other hand, belt polishing offers a superior removal rate of 15 g/min, and each belt can process three pieces. However its position control is less accurate due to its area contact and belt vibration.

Table 5 Comparison of milling, grinding and belt polishing.

| Potential process | Surface finish | Removal rate (g/min) | Cycle time (min.) | Tool life (no. of vanes) | Tool unit cost (US$) | Machine cost (US$) |
|---|---|---|---|---|---|---|
| 5-axis milling | Micron | 0.8 | >56 | 0.22 | 6 [1] | 350,000 |
| 5-axis grinding | Sub-micron | 0.15 | 300 | <3 [2] | 150 [3] | >500,000 |
| Belt polishing with 6-axis robot | Micron or Sub-micron | 15 | 2.5 | 3 | 7.5 | <300,000 |

Note:
(1) The price is for the ball end mill insert.
(2) The wheel wears faster that the material removal. It is not practical to grind more than three parts for one wheel with tool wear compensation.
(3) The price is for the conventional abrasive wheel having a size of 200 mm in diameter.

Due to the inherent difficult-to-machine nature of superalloy, any process would incur severe tool wear, with the end mill cutter being the worst. Fast tool wear means frequent tool changing and tooling cost. Again, polishing belts are much cheaper than grinding wheels, resulting in lower tool cost. Although an ultra-hard cutter would improve milling efficiency considerably and bring the cutter life close to a polishing belt, its high cost, in the range of a few thousand US dollars, hardly justifies such an automation system. Taking into account all these considerations, it can be concluded that the belt polishing process is most suitable for automating airfoil blending despite its large size and poor position controllability.

## 3. Force Control in Material Removal

In contrast to CNC milling where material removal depends on position control, abrasive machining processes like belt polishing have the contact force as the prime factor determining the material removal and final profile finish quality. Force control is particularly important for avoiding overcutting and undercutting. To achieve the desired finish profile, it is crucial to control the contact force and compliance between tool and vane, as the profile of the workpiece is rather irregular (after brazing) and distorted. An operator carries out the job skilfully with his/her vision (visual sense of distorted profile) and force feedback (muscular and tactile sense of contact force). It is a difficult task for a machine to mimic these human capabilities. Two fundamental issues must be addressed: profile sensing and fitting, and effective force control in the constrained robotic environment. Depending on the application requirements, the robotic system can be either a robot-holding-tool or robot-holding-workpiece configuration.

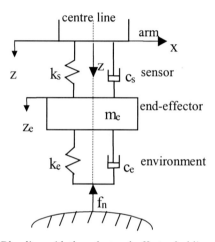

Figure 6 Blending with the robot end-effector holding the tool head.

## 3.1    *Robot Holding Tool*

The blending environment is modelled as a mass-spring-damper system, as shown in Figure 6. The force sensor is modelled as a simple spring-damper model, where $k_s$ and $c_s$ are respectively the coefficients of spring and damping of the sensor. $m_e$ is mass of the tool and robot end-effector, $k_e$ and $c_e$ are respectively the coefficients of spring and damping of contact environment, $f_s$ and $f_n$ are respectively the sensing force and vertical contact force. The following equations can be obtained:

$$f_s = k_s(z - z_e)$$
$$f_n = k_e z_e \qquad\qquad (2)$$
$$m_e z_e'' = f_s + c_s(z' - z_e') - f_n - c_e z_e'$$

where $z$ is the tip position of the arm, $z_e$ the position of the tool. Taking the Laplace transform of Equation (2), we obtain:

$$E(s) = \frac{f_n(s)}{z(s)} = \frac{k_1(1 + k_2 s)}{s^2 + 2\zeta\omega_n s + \omega_n^2} \tag{3}$$

where

$$k_1 = \frac{k_s k_e}{m_e} \qquad\qquad k_2 = \frac{c_s}{k_s} \tag{4}$$

$$\omega_n = \sqrt{\frac{k_s + k_e}{m_e}} \tag{5}$$

$$\zeta = \frac{c_s + c_e}{2\sqrt{m_e(k_s + k_e)}} \tag{6}$$

$\omega_n$ is the natural frequency and $\zeta$ is the damping ratio. When $c_s$ is negligible and not considered, the dynamics is a second-order system.

The robot-holding-tool configuration typically employs a small polishing tool while a large piece such as an aircraft fan blade over one meter long is fixed on a workstation. For a small workpiece like a jet engine turbine vane or blade polished with a large processing tool such as a wheel grinder, an alternative approach would be the robot-holding-workpiece scheme.

## 3.2    *Robot Holding Workpiece*

The robot-holding-workpiece configuration is desirable if a large amount of material is to be removed and the tool wear is severe. The approach allows the tool station to be fixed on the floor with less space constraint. Therefore, a long belt (often a few meters) can be employed. Figure 7 shows a robot holding the workpiece against the tool head. In general, there are two variations in this approach, namely fixed tool head and floating tool head.

In a fixed tool head polishing system, the contact force depends on the physical characteristics of the contact wheel and the belt, and the robot end-effector displacement. The force control can be realised by controlling the vertical movement of the workpiece against the contact wheel. However, further studies show that the force control stability cannot be easily achieved. One reason is that the wheel stiffness is not constant over the desired displacement range. When the displacement increases, the contact area between the workpiece and the contact wheel becomes larger, resulting in a higher stiffness. Tool stiffness is further affected by the workpiece condition and tool wear. Another factor is that a soft contact wheel quickly becomes a hard tool (no more deformation is possible) after a small displacement. The non-linearity and saturation effect makes the force control very ineffective. In fact, the fixed tool head approach is commonly used for hard tool material removal. The tool, such as a hard grinding wheel, is subject to pure position control as in CNC milling. To overcome the problems associated with non-linearity and stiffness saturation, the Passive Compliant Tool (PCT) concept has been proposed.

Figure 7 Robot end-effector holding workpiece.

Figure 8 illustrates the principle of Passive Compliant Tool. It consists of a tool motor, a contact wheel, one or more tension wheels, a pre-loaded spring, and a displacement sensor. The tool "sinks" when the contact force is high, but "floats" in the case of lower contact force. A long abrasive belt runs through the motor shaft and wheels. The purpose of the displacement sensor is to detect the spring extension, hence the contact force. When the robot presses down, the spring displacement exerts a greater contact force between the workpiece and the contact wheel.

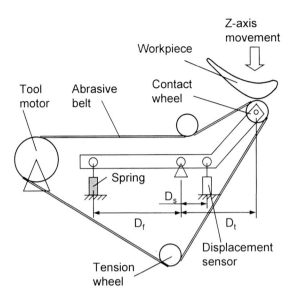

Figure 8 Blending with a floating tool head.

When the Z-axis displacement is less than the critical displacement $Z_c$, the pre-loaded spring does not extend. In other words, the PCT behaves like a fixed tool head. Beyond the critical Z-axis displacement, the spring displacement $L$, leading to the sensor measurement $S$, determines the contact force.

The following sets of equations can be obtained:

$$\left.\begin{matrix} L = 0 & (Z \leq Z_c) \\ L = R_l \, (Z - Zc) & (Z > Z_c) \end{matrix}\right\} \tag{7}$$

$$\left.\begin{matrix} S = 0 & (Z \leq Z_c) \\ S = R_s \, (Z - Zc) & (Z > Z_c) \end{matrix}\right\} \tag{8}$$

where $R_l$ is the spring displacement coefficient, and $R_s$ sensor displacement coefficient. Assuming that the distances of tool head, sensor, and the spring from the pivotal point of the beam are $D_t$, $D_s$, and $D_f$ respectively, the displacement coefficients can be obtained based on triangulation:

$$R_l = \frac{D_f}{D_t}$$

$$R_s = \frac{D_s}{D_t}$$

Assuming that the spring stiffness is $K_l$ and the initial displacement due to the pre-loading is $L_0$, the spring tension force and the contact force can be obtained as follows:

$$F_l = K_l\,(L+L_0)$$
$$F = R_l\,F_l$$
or
$$F = R_l\,K_l\,(L+L_0) \qquad (9)$$

At the critical Z-axis displacement ($L = 0$, $Z = Z_c$), the following condition is satisfied:

$$F = R_l\,K_l\,L_0 = K_t\,Z_c$$

where $K_t$ is the contact wheel stiffness coefficient. The critical Z-axis displacement can be obtained as follows:

$$Z_c = \frac{R_f\,K_l}{K_t}\,L_0 \qquad (10)$$

Combining Equation (7) and (8), we can obtain the spring displacement as a function of sensor measurement:

$$L = \frac{R_l}{R_s}\,S$$

Substituting it into the Equation (9), the contact force can be obtained as a function of sensor displacement:

$$F = R_l\,K_l\,L_0 + \frac{R_l^2\,K_l}{R_s}\,S \qquad (11)$$

Equation (8) and (11) govern the force control. Figure 9 shows the sensor displacement and contact force as the functions of Z-axis displacement. In an unknown constrained blending environment, the desirable contact force can be achieved through active force control. The displacement sensor measures the contact force based on Equation (11), the Z-axis displacement is controlled in real time according to Equation (8).

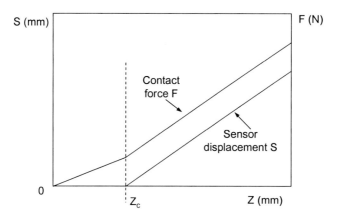

Figure 9 Response of contact force to Z-axis movement.

Given the geometry of the workpiece, a desirable force level can be achieved by associating the robot Z-axis offsets with all tool path points. Once tooling conditions including pre-load, spring stiffness, contact wheel hardness and construction, are known, the pre-determination of Z-axis offsets is sufficient to maintain the desirable force level. The Z-axis offsets can be easily incorporated into adaptive robot path and strategy planner. Such a control scheme works well to compensate for global variations, mainly due to part distortions. To compensate for local variations such as variable braze thickness and transitional lines from non-brazing area to brazing area, sensitivity of the contact force to brazed layer variation can be adjusted to the required level by changing the spring stiffness and the pre-load. From numerous laboratory tests, an optimum polishing operational model is established. The process knowledge is encapsulated so that optimum process parameters can be inferred according to individual part conditions.

## 4.     Model-Based Robotic Machining

Robots have been increasingly used in machining applications such as cutting, deburring, chamfering, grinding and polishing. By and large, the existing systems are model-based [20, 21]. In the context of computer integrated, flexible and automated manufacturing, robots play an important role as a flexible component in automation. Computer-aided tools for robotic modelling, task planning and simulation, programming, and sensor integration have been vigorously researched. In recent years, there has been an explosion of research interests in robot task planning, off-line programming, automatic program synthesis, and robot integration into manufacturing system. Chen et al [22] discussed an off-line programming system based on CATIA software.

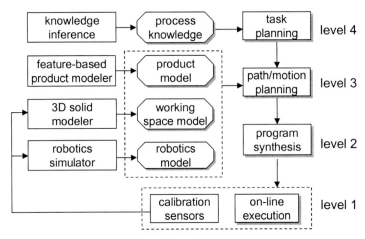

Figure 10 Model-based robotic system.

A model-based robotic system can be generalised in Figure 10. It consists of four task levels:

- **Level 1 – on-line execution**. At the lowest level, the machine controller accepts the synthesised program, and executes the machining tasks accordingly.
- **Level 2 – program synthesis**. The neutral files are synthesised, and translated into programs that are accepted by a target machine. In a CNC system, NC codes are generated by a postprocessor. In robotics application, proprietary tools are available to generate programs in a target robot language. The target machine program can be further verified in the virtual machining environment before being downloaded into the machine controller.
- **Level 3 – path/motion planning**. Once the task specification is generated, a designer can plan the robot path and motion in a computer environment. To fulfil the task, the designer needs the product model, robotics model, and workspace model that can be readily built with CAD and robotics simulation tools. The output of this level is still CAD data, often called neutral files.

- **Level 4 – task planning**. At the highest task level, a designer generates machining sequence and tooling requirements based on expert knowledge. Researchers are working on intelligent knowledge processing tools to generate optimum task specifications automatically.

A model-based robotic machining system offers the following advantages:

- To facilitate the integration of robotic machining systems into a CIM environment.
- To minimise or eliminate robot down time and to shorten production time, hence to reduce production costs.
- To overcome the disadvantages of on-line manual programming which is time-consuming and tedious.
- To support planning engineers for efficient robotic task planning and analysis, graphical simulation, calculation of cycle time, collision detection, and program generation.
- To integrate calibration sensors into robotic systems so that robots can be more adaptable to the actual environment.

One of the challenges in applying model-based robotic systems is the discrepancy between the ideal world (computer model) and the real world. Robotics practitioners are resorting to calibration means to overcome the problem. Sophisticated tools are available to calibrate robot kinematics. An example is robot pose performance test using proximity sensors [23]. Calibration sensors can be employed in the real world to acquire faithful information about the workspace and workpiece to be handled, as shown in Figure 10. These data are fed back to the computer models for corrections.

A calibration approach is commonly used in robot set-up, re-installation, or changing of product types to solve static discrepancies. However, it is ineffective in dealing with process dynamics and part-to-part variations. The dynamic variations render real-time sensory control and knowledge-based control necessary [24, 25].

## 5.    Part Variations and Process Dynamics

Jet engine turbine components to be overhauled are often severely distorted and twisted during their service in a high-temperature and high-pressure environment. It is extremely difficult to remove the excess material and polish the brazed vane, (either manually or using a robot) so that the prior-to-braze airfoil can be accurately reconstructed. Apart from the its poor machinability, each component has a different level of geometrical, positional and dimensional variability in the following areas:

- Tooling hole dimensions, tolerances and position.
- Leading edge dimensions, tolerances and position.
- Trailing edge dimensions, tolerances and position.
- Location of the transition lines on brazed patterns.
- Overall thickness of vane profile.
- Amount of excess material.
- Variable curvature ratio of the airfoil.

In addition, each airfoil and braze pattern may require a separate set of optimum process parameters that in turn become a significant source of potential errors in an automated blending system. These process parameters include the following items:

- Grit size, type of bonding of the grit, tension of the polishing belt.
- Wheel material, diameter, construction, stiffness, hardness, and contact angle.
- Contact wheel deformation.
- Belt speed, vibration, and wear.
- Tool head vibration.
- Approach angle.
- Feed rate.
- Contact force.
- Removal rate.

The sources of errors, system variables and proposed solutions are summarised in Table 6. Unlike a conventional CNC machining system, a small error or change in one of the system variables will adversely affect the resultant accuracy of the turbine airfoil. Slight errors, for example, in the orientation of the turbine airfoil, trajectory of the robot or the position of the grinding wheel will create a large compounded angular error on the grinding wheel contact angle and its approach angle. The grinding wheel contact angle and its approach angle will in turn create an undesirable dwelling on the airfoil curvature if such an error is not adequately compensated.

Table 6 Summary of sources of errors, system variables & solutions.

| Sources of errors | Variables | Est. errors | Proposed solutions |
|---|---|---|---|
| | **Part variations** | | |
| Part distortions | Dimensions, position of trailing edge, curvature ratio | 1.5 mm | Measurement & s/w compensation |
| Leading edge | Position of LE in reference to the buttress | 1.5 mm | Measurement & s/w compensation |
| Brazing pattern | Location of the transition lines, and braze thickness | 0.5 mm | Tool path optimisation |
| | **Robotics & tooling errors** | | |
| Robot repeatability | Six coordinates in the global coordinate system (CS) | 0.1 mm | Layout design optimisation |
| Robot end-effector | Six coordinates in the robot hand CS | 0.01 mm | Self-aligned end-effector. |
| Tooling position | Six coordinates in the tool CS | 0.1 mm | Intuitive tool calibration |
| Profile measurement | Sensory measurements | 0.01 mm | Inherent, cannot be improved |
| | **Process parameters** | | |
| Wheel construction | Contact wheel material, diameter, hardness, construction, contact angle | Empirical | Process optimisation |
| Belt construction | Grits, bonding of grits, belt tension | Empirical | Process optimisation |
| Polishing belt wear | Changes in removal rate | Empirical | Automatic tool wear compensation |
| Process dynamics | Feed rate, contact force, belt speed & vibration, belt wear, tool head vibration, tool approaching angle | Empirical | Knowledge-based process control |

From the industrial perspective, an automated 3D blending system must be able to overcome part variations and process dynamics with minimal human intervention. Therefore, to meet the quality requirements specified in Table 3, the following automation capabilities are required:

- Capability of detecting deviations of up to 2.0 mm of the actual profile from the design profile.
- Automatic compensation for the part distortions to avoid overcutting.
- Non-interrupted robot operation during manual tool changing
- Sufficiently long tool life
- Automatic tool wear compensation.
- Significant improvement of cycle time, as compared with 10 minutes for manual polishing.
- Flexibility in expanding to other configurations of the same vane or other products.

The part variations, largely geometric attributes, can be captured through an in-situ measurement system. Intelligent software compensation is capable of overcoming the variations. However, process parameters are empirical, and cannot be established in an analytical model. A new system concept has to be explored.

## 6.    System Concept of Adaptive Robotic Blending System

### 6.1    *A Mechatronic Approach*

Current robot technologies cannot handle the part distortions and process dynamics inherent in component overhauling due to the following limitations:

- Conventional industrial robots are mainly for repetitive simple tasks.
- Finishing robots have been used in the areas such as polishing musical instruments, aircraft manufacturing parts, deburring, but largely based on teach-and-play.
- Off-line programming systems, such as IGRIP, RobCad, CimStation, Grasp and Workspace, are based on design data and calibration data.
- Real-time sensory control has been practiced in non-constrained material processing, such as spraying, laser processing and welding to accommodate part variations. But constrained material processing like blending involves much more complicated process dynamics.

It is then decided that a mechatronic approach has to be innovated to meet the challenges [26]. On top of a suitable industrial robot that meets the rigidity, dexterity and repeatability requirements, intelligent automation capabilities must be developed around it. Solutions to part variations and process dynamics lie in:

- In-situ Profile Measurement (IPM), which acquires seed points on individual airfoils.
- Optimal Profile Fitting (OPF), which finds the airfoil geometry through mapping between the design data (as templates) and the actual measurement points.
- Adaptive Robot Path Planning (ARP), automatically computes the tool path based on the true profile, and generates the robot program.
- Knowledge-Based Process Control (KBPC). It automatically compensates tool wear to achieve constant material removal, and optimises the tool path to achieve the required finish quality.

Figure 11 shows the architecture of the Knowledge-Based Adaptive Robotic System for 3D profile blending. It comprises three inter-related hierarchical layers, namely, Device and Process, Knowledge-Based Process Control (KBPC), and Data-Driven Supervisory Control (DDSC). Within each layer there are several control modules.

## 6.2    *Device and Process*

Device includes a finishing robot and servo-driven Self-Aligned End-Effector (SAE), Passive Compliant Tool (PCT) heads and an index table. In addition, it includes various sensors as follows:

- In-situ Profile Measurement (IPM) sensor.
- Pneumatic sensor for detecting part jamming in the end-effector.
- Laser through-beam sensor for confirming proper gripping.
- Inter-lock sensor for safety measure.
- Belt breaking detector.

Rough grinding is designed to remove the bulk of the braze material from the airfoil, but still leaves some transitional lines on the airfoil. The fine polishing process removes all transitional lines and overlapping marks, and polishes the profile to the final surface roughness. The rationalised process eliminates the fine polishing with flipper wheel as in the manual process. The streamlining also simplifies the system design, furthermore reduces the cycle time.

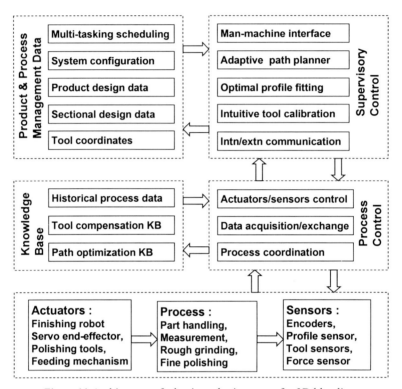

Figure 11 Architecture of adaptive robotic system for 3D blending.

## 6.3    *Knowledge-Based Process Control (KBPC)*

The Process Control sub-system controls actuators and sensors to fulfil the required processes, and coordinates the process flow. In addition, it also acquires measurement data and exchanges data and information with the Supervisory Controller. Process Control relies on the following process knowledge bases:

- **Historical Process Database**. It holds records of individual parts, such as measurement data, processing time, measurement data of finished profile (optional), breakdowns, uptime, and downtime.
- **Tool Compensation Knowledge Base**. It stores tool compensation parameters such as the abrasive belt speed and the workpiece feed rate.

- **Path Optimisation Knowledge Base**. It contains optimum process parameters such as Z-axis offset, approaching angle, robot travel speed. Optimum parameters can be inferred based on the part conditions: curvature, braze thickness, leading edge height, and trailing edge distortions. As a result, a smooth finish profile, free of transition lines, overlapping marks and burning marks, can be obtained.

## 6.4    *Data-Driven Supervisory Control (DDSC)*

Product and process management data such as multi-tasking scheduling, system configuration, product design data, sectional profile data, and tool coordinates drive the Supervisory Controller. The following control functions were implemented in the sub-system:

- **Internal / External Communication**. Internal communication involves information and data exchange between control modules in DDSC. External communication allows data exchange with KBPC.
- **Intuitive Tool Calibration (ITC)**. During machine re-calibration, set-up or re-installation, position data of tool stations, measurement stations, and index table can be manually clocked. These data are keyed into the database. The mathematical model of workspace and robot kinematics is automatically generated.
- **Optimal Profile Fitting (OPF)**. It generates the actual profile based on the in-situ measurement data. The robust fitting algorithm uses the sectional data as templates, and maps them with the measurement points. 3D free form surface is generated through interpolation of cross sectional profiles.
- **Adaptive Path/Strategy Planner (ARP)**. It automatically generates the optimum tool path based on individual part conditions, and furthermore synthesises the robot programs from the resultant path points.
- **Human-Machine Interface (HMI)**. It allows the user to change some system parameters, configure system, select product configurations, enter data, and make queries.

## 6.5    *System Layout and Working Cycle*

The conceptual layout of the adaptive robotic blending system, shown in Figure 12, consists of a 6-axis finishing robot with a payload of at least 30 kg, four PCTs, a workpiece feeding unit, and an in-situ profile measurement unit. Tool #1 (rough polishing) and #2 (fine polishing) are grouped in one site and #3 (rough polishing) and #4 (fine polishing) in the mirrored site. This split-envelope configuration allows the operator to change the belt safely in one site while the robot is working in another site, thus minimising the robot idle time.

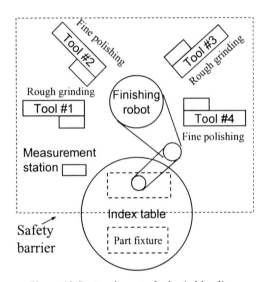

Figure 12 System layout of robotic blending.

A rationalised working cycle of the robotic grinding/polishing process is shown in Figure 13. To start a cycle, the finishing robot picks up a workpiece from the feeding unit. It then performs in-situ measurement on selected points of the airfoil profile. The measurement data of the airfoil is sent to the host computer. By using those measurement data and the design airfoil profile, the template-based optimal profile fitting is performed. Based on the fitted profile, robotic grinding/polishing paths are computed and downloaded to the robot controller. After receiving the grinding/polishing paths, the finishing robot carries the workpiece and performs the specified grinding/polishing process on the grinding/polishing stations. The finished workpiece is finally returned to the index table. The robot then picks up a new workpiece and starts another working cycle.

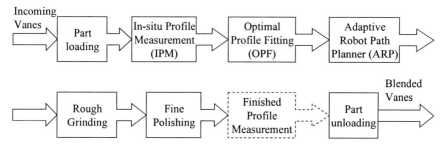

Figure 13 Operational cycle of the proposed adaptive robotic blending system.

## 7.    Process Optimisation

Grinding and polishing a superalloy material like Inconel is a very complex process, involving many process variables. Desired finish profile and quality can only be achieved with optimum process parameters, tool path and tool wear compensation. The relationships between these variables and finish requirements are not analytical, but rather empirical. In our approach, all this human-intrinsic process knowledge is encapsulated in the knowledge base, some of which can be obtained from human operators. Much more knowledge has to be obtained from extensive laboratory prototyping [27].

### 7.1    *Grinding/Polishing Process Parameters*

*Belt Linear Speed*

It is proportional to the rotational speed of the motor that drives the belt. The maximum rotation speed is 3000 rpm. The recommended speed is between 700 rpm to 1800 rpm, considering the belt stabilisation.

*Feed Rate*

It is the tangential speed at which the workpiece moves against the tool. The feed rate can be fixed in a part program. The robot system also provides a feed ratio or factor for users. The actual feed rate is thus the product of the ratio and the pre-determined moving speed of the robot hand.

*Depth of Cut*

It is determined by the offset value of the robot hand along the $z$-axis.

*Pre-Load*

It is the spring force applied on the workpiece. The load can be adjusted by changing the spring coefficient of and its elongated length.

*Abrasive Belt*

It is the tool used for grinding and polishing turbine vanes in this project. The commonly used conventional abrasive belts are Norton and 3M. The mesh size of belts used for engine components overhaul is normally between 40 and 120.

The abrasive material of the belts used for PCTs is ceramic aluminium oxide with a mesh size of 80. The abrasive is specially designed for grinding aerospace alloys and forged steels. The braze material to be ground is an alloy containing about 45% of cobalt, 24% of chromium, 25% of nickel and 3.5% of tungsten. It falls into the category of hard-to-grind materials. The belt is used for both rough grinding and fine polishing. The fine polishing tool differs from rough grinding tool in two aspects:

- The rough grinding tool uses a stiff spring and high pre-load to have a stiffer contact between the tool and workpiece. A polishing tool head utilises a less stiff spring and less pre-load. Therefore, the process parameters in rough grinding are very different from those in fine polishing.
- The rough grinding tool uses a brand new belt to achieve the maximum removal rate. On the other hand, the belt for fine polishing was a used belt from the grinding process. The used belt has a lower material removal rate, but produces better surface finish. With this approach, not only can we meet the surface finish requirements, but also significantly reduce the production cost and simplify tool management.

## 7.2    *Tool Path Optimisation*

The passive compliance tool combined with adaptive path generation enables the achievement of a good approximation of the required surface quality and dimensional accuracy, but is insufficient to achieve the final finish requirements. To achieve the required profile smoothness, the robot polishing strategy must be planned according to the conditions of a specific transition line. The robot trajectory must be tuned to remove transition lines between brazed and non-brazed areas, yet not over-cut the non-brazed surface.

Figure 14 illustrates two examples of removing transition lines on concave and convex sides respectively. Tool paths were slightly modified along the arrowed lines, instead of following the vane contour as planned. Four repeated cuts were planned to remove these transition lines before grinding the whole profile. The approach to and departure from the transitions lines at certain angles emulate the operator's capability of wrist adjusting to achieve smooth curvatures.

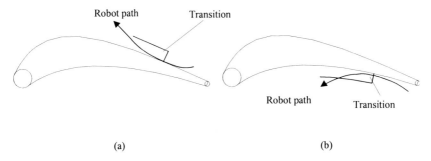

(a)                                                        (b)

Figure 14 Path optimisation for removing transverse transition lines.

Another difficulty in achieving smooth profiles is how to avoid overlapping lines between two parallel cuts. Cutting paths based on conventional CAM definitely leaves visible lines in the boundary areas. Our solution is to adjust tool path angles and optimally overlap two parallel cutting passes, resulting in a smooth transition from one cutting pass to another. With the path and trajectory optimisation, based on the process knowledge, smooth airfoil profiles were achieved, free of transition lines and overlapping lines.

In the case of belt grinding with passive force control, the material removal volume $V_m$ is mainly dependent on the contact force between the tool and the workpiece, the belt speed and the feed rate along the tangential direction. The effects of these three parameters on the material removal rate were investigated experimentally. Results are shown in Figure 15. The amount of material removal is normalised against the vane weight, to ensure that the comparison is done on an equal basis. It is apparent that the normalised material removal amount increases as the contact load and belt speed increase, but as the feed rate decreases.

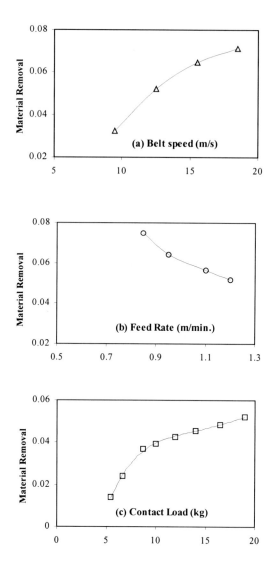

Figure 15 Effect of (a) belt speed for feed rate 1.2 m/min & load 19 kg,
(b) feed rate for belt speed 12.5 m/s & load 19 kg,
(c) contact load for belt speed 12.5 m/s & feed rate 1.2 m/min.

## 7.3   *Tool Wear Compensation*

When using the abrasive belt to grind turbine vanes, tool wear is significant. As shown in Figure 16, the normalised removal amount decreases with the progress of grinding (in terms of vane numbers being ground) for one abrasive belt. After grinding three vanes, the material removal decreases substantially. Scanning Electron Microscope (SEM) examination showed that the wear of abrasive grains was developed after grinding one vane and progressively worsened with further grinding, as illustrated in Figure 17. Grain pulling-off was often observed on belts that had ground three vanes, as clearly seen in Figure 17 (d), corresponding to the abrupt drop of the material removal amount in Figure 18. This is apparently a result of losing some effective cutting grains.

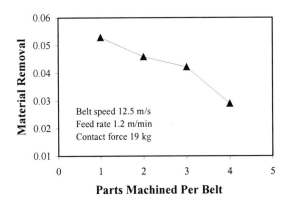

Figure 16 Effect of belt wear status on material removal.

For an automatic grinding and polishing system, the maintenance of a reasonably constant material removal rate is important. Compensation must be made for tool wear. The decrease of the material removal rate due to belt wear can be compensated for by increasing the abrasive belt speed and decreasing the workpiece feed rate, according to the results shown in Figure 16. The result is satisfactory.

As can be seen in Figure 18, the normalised material removal after tool wear compensation (empty squares) is much more consistent compared with that before compensation (solid squares). Similarly, tool wear compensation results in more consistent surface finish roughness. However, there is only a limited range over which tool wear compensation can be applied by increasing the belt speed and decreasing the feed rate, as high belt speed and low feed rate can cause burn marks on the airfoil surface. Such defects do not meet the quality requirements. For rough grinding the belt must be replaced after grinding three vanes.

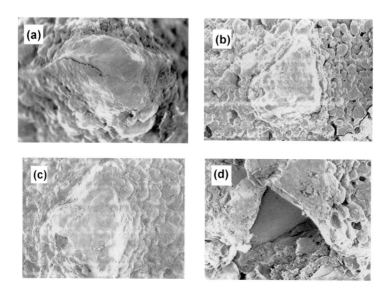

Figure 17 Wear status of an abrasive belt (a) before grinding, (b) after grinding 1 vane, (c) after grinding 2 vanes and (d) after grinding 3 vanes.

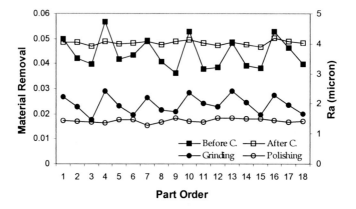

Figure 18 Comparisons of material removal (squares) before and after tool wear compensation and surface finish (circles) after grinding and polishing.

The process model derived from the expert knowledge and process optimisation experiments drive the Knowledge-Based Process Control (KBPC). It infers the optimum process parameters according to the part and process conditions, and hence compensates errors arising from variable belt conditions, variable contact wheel conditions, and process dynamics.

## 8.     Concluding Remarks

Superalloys such as Inconel fall into the difficult-to-machine category. The current manual blending operation, using belt polishing tools, is very labour-intensive, and requires extensive process knowledge and skills. Unfortunately, no other alternative process could provide better automation solutions. Despite poor position controllability in the belt polishing, our automation solution has to be built around such an artefact-type process, and mimic the manual operation to a large extent.

Force control determines the material removal as opposed to position control in the case of CNC milling. The model of the contact force between the tool and the workpiece has been derived. Both active force control and passive force control can be implemented, depending on the blending requirements. In the specific application, related to turbine vane blending, passive force control meets the material removal control requirements and can be readily implemented, hence the control scheme is recommended for the final implementation.

Part variations and process dynamics, predominant in the robotic 3D blending, are recognised as the major causes of concern. Model-based robotic systems cannot deal with them effectively. A knowledge-based adaptive robotic system integrating all intelligent control modules, Knowledge-Based Process Control (KBPC) and Data-Driven Supervisory Control (DDSC) provides an effective solution to the practical industrial automation problem. Process optimisation has been conducted through expert knowledge encapsulation and numerous laboratory tests. The resultant empirical process model drives Knowledge Based Process Control (KBPC). Optimum process parameters are inferred from the knowledge bases according to part and process conditions, and hence compensates errors arising from variable tooling conditions and process dynamics. Implementation of the full system will be discussed in Chapter 3.

# References

1.  Whitney, D.E. "Historical perspective and state of the art in robot force control," The International Journal of Robotics Research, 1987, Vol.6, No.1, pp. 3-14.

2.  Hogan, N. "Impedance control: an approach to manipulation – Part I: Theory; Part II: Implementation; Part III: Applications", Transaction of ASME Journal of Dynamic Systems, Measurement and Control, 1995, Vol. 107, No. 1, pp. 1-24.

3.  Raibert, M.H., and Craig, J.J. "Hybrid position/force control of manipulator", Transaction of ASME Journal of Dynamics Systems, Measurement and Control, 1981, Vol.102, pp. 126-133.

4.  McClamroch, N.H., and Wang, D. "Feedback stablization and tracking of constrained robots", IEEE Transaction on Automatic Control, 1988, Vol. 33, No. 5, pp. 419-426.

5.  Ge, S.S., Hang, C.C., Woon, L.C., and Chen, X.Q. "Impedance control of robot manipulators using adaptive neural networks", International Journal of Intelligent Control and Systems (World Scientific, USA), 1999, Vol. 2, No. 3, pp. 433-452.

6.  Lucibello, P. "A learning algorithm for hybrid force control of robot arms", 1993 IEEE International Conference on Robotics and Automation (IEEE Robotics and Automation Society), Atlanta, 1993, pp. 654-658.

7.  Suzuki, T., et al. "Contact force control for 2 D.O.F. manipulators based on iterative learning operation," Proceedings of the 1992 International Conference on Industrial Electronics, Control, Instrumentation and Automation (The Industrial Electronics Society of the IEEE, The Society of Instrument and Control Engineers of Japan), San Diego, 1992, pp. 682-687.

8.  Arimoto, S., and Naniwa, T. "Learning control for motions under geometric endpoint constraint," 1992 American Control Conference, Chicago, 1992, pp. 2634-2638.

9.  Jeon, D., and Tomizuka, M. "Learning hybrid force and position control of robot manipulator", IEEE Transaction on Robotics and Automation, 1993, Vol.9, No.4, pp. 423-431.

10. Wang, D., Soh, Y.C., and Cheah, C.C. "Robust motion and force control of constrained manipulators by learning," Automatica, 1995, Vol. 31, No. 2, pp. 252 – 262.

11. Chen, X.Q., Wu, Y.F., Fong, A.M., and Zhu, J.Y. "Successive learning reinforcement to overcome dynamic unpredictability in material surface blending", Proceedings of the Forth International Conference on Control,

Automation, Robotics and Vision, Singapore, 3-6 December 1996, pp. 609-613.

12. Guglielmo, K., and Sadegh, N. "Implementing a hybrid learning force control scheme," IEEE Control System, 1994, Vol.14, No.1, pp. 72-79.

13. Stephien, T.M., Sweet, M.L., Good, M.C., and Tomizuka, M. "Control of tool/workpiece contact force with application to robotic deburring", IEEE Journal of Robotics and Automation, February 1987, Vol. RA-3, No. 1, pp. 7 – 18.

14. Proctor, F.M., and Murphy, K.N. "Advanced deburring system technology", ASME, PED-Vol. 38, 1989, pp. 1-12.

15. Bone, G.M., Elbestawi, M.A., Lingarkar, R. and Liu, L. "Force control for robotic deburring", ASME Journal of Dynamic Systems, Measurement and Control, 1991, Vol. 113, pp. 395-400.

16. Elbestawi, M.A., Bone, G.M. and Tam, P.W. "An automated planning, control and inspection system for robotic deburring", Annals of CIRP, Vol. 41/1/1992, pp. 397-401.

17. Hollowell, R., and Guile, R. "An analysis of robotic chamfering and deburring", ASME, DSC-Vol. 6, 1987, pp. 73-79.

18. Nasu, Y., Mitobe, K., Kouda, N., Wakabayashi, K., and Honda, Y. "Robotic deburring of GFRP parts", The Second International Conference on Control, Automation, Robotics and Vision, Singapore 2-6 December 1992, pp. RO-16.1.1-RO-16.1.5.

19. Ge, D.F., Takeuchi, Y., and Asakawa, N. "Dexterous polishing of overhanging sculptured surfaces with a 6-Axis control robot," Proceedings of IEEE International Conference on Robotics and Automation, 1995, pp. 2090-2095.

20. Chen, X.Q., Gay, R., & Lim, R. "Design and implementation of an integrated shop floor control system", Proceedings of the Third International Conference on Automation, Robotics, and Computer Vision (Vol. 2), 8-11 November 1994, Singapore, pp. 793-797.

21. Zha, X.F., Lim, S.Y.E., Fok S.C. and Chen, X.Q. "CAD-driven hybrid intelligent strategies generation for robotic assembly task plan execution and control", Proceedings of the Fifth International Conference on Control, Automation, Robotics and Computer Vision, Singapore 9-11 December 1998, pp. 1492-1497.

22. Ye, N., and Chen, X.Q. "Robot pose performance and related test equipment", Industrial Automation Journal, January – March 1995, pp. 14-17 & 30-31.

23. Chen, X.Q., Lim, B.S., and Lim, R. "An integrated approach towards robotic modular fixture assembly", Proceedings of the Third International Symposium

on Measurement and Control in Robotics, Torino, Italy, 21-24 September 1993, pp. Bm.I-25 to Bm.I-30.

24. Mori, K., Kasashima, N., and Yamane, T. "Real time knowledge based control for machining," Modern Tools for Manufacturing Systems, 1993, pp. 69 – 76.

25. Inasaki, I. "Sensor fusion for monitoring and controlling grinding processes", International Journal of Advanced Manufacturing Technology (1999), pp. 15:730-736.

26. Chen, X.Q., Gong, Z.M., Huang, H., Zhou, L.B., Ge, S.S., Zhu, Q. and Woon, L.C. "An automated 3D polishing robotic system for repairing turbine airfoil", Proceedings of 3rd International Conference on Industrial Automation, Canada, June 7-9, 1999, pp. 11.9 - 11.13.

27. Huang, H., Gong, Z.M., Chen, X.Q., and Zhou, L.B. "Robotic grinding/polishing for turbine vane overhaul", International Conference on Precision Engineering (ICoPE2000), Singapore, 21-23 March 2000, pp. 582-587.

# CHAPTER 3

# ADAPTIVE ROBOTIC SYSTEM FOR 3D PROFILE GRINDING/POLISHING

XiaoQi Chen*, Zhiming Gong*, Han Huang*, Shuzhi Ge**, Libo Zhou***

*Gintic Institute of Manufacturing Technology,
71 Nanyang Drive, Singapore 638075*

**Department of Electrical & Computer Engineering, The National University
of Singapore, 10 Kent Ridge Crescent, Singapore 119260*

*** Department of System Engineering, Ibaraki University, Japan*

## 1.    Introduction

Robotic machining has certain advantages over conventional CNC machining: high flexibility, capability of integration with peripherals such as sensors and external actuators, and lower cost. Attempts have been made to apply robotic machining to advanced material processing. United Technology Research Centre (UTRC) has been developing robotic machining technologies since early 80's [1]. A two-axis closed-loop controlled micro-manipulator has been developed for automated chamfering and deburring, but is yet to be further explored for polishing 3D profiles. Berger et al [2] have reported an advanced mechatronic system for turbine blade manufacturing and repair. Their work proposed a high-speed multi-axis milling machine to cut difficult-to-machine materials. A 3D profile measurement sensor system was proposed for compensating the part tolerances. Again the work was confined to a laboratory exploration.

Most recently, Chen and Hu [3, 4] have implemented a robot system for sculpture surface cutting, meant for rapid prototyping. A part model is used to generate robot path and trajectory. The 3D sculpture surface is produced based on robot position control. In [5], a force controlled robotic finishing system is discussed. The idea is that a force-controlled robot can follow the edges or the surfaces of the workpiece using force control functions. Kunida and Nakagawa [6] have developed a curved surface polishing robot system using a magneto-pressed tool and a magnetic force sensor. Desired contact force can be maintained during polishing. In more sophisticated robotic machining applications, dual manipulators may be required. Adaptive neural network control has been attempted for coordinated manipulation in a constrained environment [7]. As far as practical applications are concerned, robotic machining has been mostly restricted to simple operations under well-defined conditions, such as deburring, polishing and chamfering of new parts.

As compared with manufacturing new parts, one of the major difficulties in overhauling aerospace components is that the part geometry is severely distorted after service in the high-temperature and high-pressure condition. As a result, the intended automation system cannot rely on the teach-and-play or programming-and-cut methods used for conventional robotic or machining applications. Another challenge is to overcome the process dynamics that is very much empirical and largely knowledge-based. Despite extensive research work and laboratory prototyping and implementation by researchers all over the world, automated systems for blending and polishing of 3D distorted profiles, such as refurbishment High-Pressure Turbine (HPT) vanes, do not exist in today's factories. The operation is manually done in almost every overhaul service factory. To this end, a concerted effort has been made to implement a robotic system for 3D profile grinding and polishing for production use.

Following the discussion on the perspective and approach of 3D profile grinding and polishing in Chapter 2, this chapter focuses on the development of core technological modules and full implementation of a working prototype. Section 2 discusses the selected finishing robot, Self-Aligned End-Effector (SAE), and control interface. Section 3 explains the In-Situ Profile Measurement (IPM) and coordinate transformation necessary to construct the part geometry. Template-based Optimal Profile Fitting (OPF) requirements, algorithm, and software development are detailed in Section 4. It is followed by the discussion on Adaptive Robotic Path Planning (ARP) in Section 5. Section 6 highlights the working prototype "SMART 3D Grinding/Polishing System", the first-of-its-kind for blending distorted 3D profiles. Section 7 presents the results of benchmarking tests conducted on the SMART system for JT9D High-

Pressure Turbine (HPT) vanes. Finally, the chapter is concluded with some remarks on the technological breakthrough and its implications.

## 2. Finishing Robot and Control Interface

### 2.1 *Finishing Robot*

In order for a line tool (contact area between the contact wheel and the workpiece can be approximated to a line) to follow a 3D surface profile, the finishing robot must have a greater dexterity as compared with a dedicated 5-axis CNC machine tool. It is quite natural that the operator uses two hands to manipulate the part to obtain desired contact force and compliance between the part and the tool. In principle, two robot arms could imitate the operator's two arms to achieve the required dexterity and rigidity. However, it poses tremendous difficulties in controlling the two mechanical arms to accomplish contact tasks. Instead, a 6-axis robot arm is preferred for its ease of motion control.

A robot for 3D blending operations has to overcome extreme reactive forces as compared to conventional industrial robots for welding, pick-and-place, glue dispensing and painting. The finishing robot not only has to hold the turbine vane in position, but also to press the airfoil surface against the grinding wheel in the normal direction and with controlled contact force, in order to achieve the desired material removal. Thus a finishing robot must have a high loading capacity, stiffness, rigidity and desired dynamic performance when used for airfoil blending operation. After an in-depth evaluation, the six-degree-of-freedom Yamaha Z-II6 robot was chosen for the blending application, shown in Figure 1. Its $\theta$, $R$ and $Z$ axes determine the position of the Tool-Centre-Point (TCP), i.e., $(X, Y, Z)$ coordinates, while the $\alpha$, $\beta$, and $\gamma$ axes, which form a Roll-Pitch-Roll wrist configuration, determine the orientation of the tool frame. The robot can carry a payload as high as 40 Kg and has a repeatability of 0.1 mm in the worst case. The robot, driven by AC servomotors with absolute position sensing, is dust-proof by applying a positive internal pressure, which is advantageous in the dusty blending environment.

### 2.2 *Self-Aligned End Effector*

In order to cater to both concave and convex airfoils, a servo-driven Self-Aligned End-Effector (SAE) has been developed, as shown in Figure 2. It has an active servo drive mechanism at one end (left) and a passive follower at the other end (right). The active end is mechanically coupled to the robot end-axis γ which has a driving torque about 35 N•m. The servo-driven SAE can rotate 360 degrees so that both concave and convex airfoils can contact the grinding wheel of the belt polishing machine in normal directions.

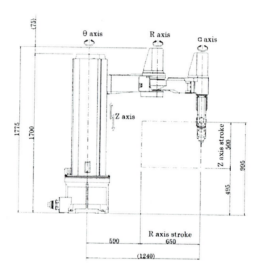

Figure 1 Yamaha six-axis finishing robot Z-II6.

Figure 2 Servo-driven self-aligned end-effector.

The passive follower ensures that the vane is held in place firmly, and in the meantime secures the axial alignment of the vane. In addition, there are three locators in the SAE to align the vane in a fixed direction. The innovative design minimises the gripping inaccuracies that would otherwise compound the airfoil distortions. The SAE has incorporated pneumatic sensors to detect any part jamming in SAE. An external laser through-beam sensor is integrated to the feeding table to check improper gripping.

## 2.3    *Control Interface*

As discussed in Chapter 2, there are two sub control systems, namely the Knowledge-Based Process Controller (KBPC) and the Data-Driven Supervisory Controller. The former, controlling all actuators and sensors, is implemented into the industrial robot controller with a powerful robot programming language dedicated to the blending operation. The latter is implemented into the host computer running the Windows NT operating system. Figure 3 shows the system communication and interface between the robot controller, interface PC, and host PC.

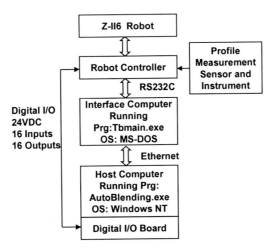

Figure 3 System communication and interface.

The host computer is interfaced to the robot controller through digital I/Os for handshaking, but the data transactions between the two controllers become difficult. The chief reason is that the robot controller can only communicate with an interface computer, running the manufacturer's program "Tbmain.exe" in MS-DOS, through a RS-232 serial line in the manufacturer's proprietary protocol. To bypass the problem, the host computer is connected to the interface PC via an Ethernet LAN. Data transactions between robot controller and host controller are relayed by the interface PC. The host computer runs the specially developed main program "AutoBlending.exe".

Via the interface computer, measurement data are sent from the robot controller to the host computer in the form of text files through a RS232C interface and an Ethernet network. The computed blending paths are sent back from the host computer to the robot controller by the same route. The executions of the robot programs and the host computer programs are synchronised by the digital inputs/outputs between them.

## 3.    In-Situ Profile Measurement

### 3.1    *Off-Line versus In-Situ Approach*

The turbine airfoils to be repaired have severe distortions and twists after operations in the high-temperature and high-pressure environment. A teach-and-play robot cannot cope with the distorted profile, nor can a commercial off-line programming system which generates a robot path according to the design data. Due to severe part distortions and part-to-part variations of the turbine vanes for repair, the design (nominal) profile cannot be used directly in a robotic surface finishing process. It is absolutely critical and necessary to have a profile sampling and distortion compensation system in this specific application to deal with part-to-part variations. Before any distortion compensation, the actual profile has to be sampled. Individual robot paths can then be generated for each workpiece from its distorted airfoil profile, which can only be obtained through measurement and profile fitting.

Two approaches of profile measurement have been evaluated: Off-line Profile Measurement (OPM) and In-Situ Profile Measurement (IPM). The former utilises an external instrument to measure the profile on a separate fixture. Then the workpiece is released from the fixture, and transferred to the robotic blending system. In the meantime, the measurement data are transmitted to the Supervisory Controller for further processing. The advantages of OPM are:

- Accurate measurement can be obtained.

- The measurement can be carried out in parallel with robotic blending operation, hence shortening the cycle time.

However, the Off-line Profile Measurement approach suffers some setbacks.

- A separate measurement station incurs extra costs. As a minimum configuration, there should be a XYZ table carrying the probe (a commercial CMM for example), and one-axis rotary table to rotate the workpiece.
- Changing fixturing from the measurement station (held by a rotary table) to the blending system (held by the robot end-effector) introduces datum errors that erode the seemingly accurate measurements obtained.

On the other hand, the In-Situ Profile Measurement (IPM) method utilises the robot itself as the measurement instrument together with a range or displacement sensor. Although the robot accuracy (0.1 mm) is much worse than a CMM (about 5 microns), the same end-effector for both profile measurement and the blending operation ensures a common datum, hence minimising fixture errors. Furthermore, the sensor head can be placed in an area where the robot has a better repeatability of less than 50 microns.

## 3.2    *Sensor Techniques*

Profile measurement sensors fall into two broad categories: contact and non-contact sensors. In the latter, an optical sensor can be used to measure 3D profile. The range sensing principle is based on triangulation. This approach has been evaluated with a 3-axis XYZ table carrying a laser range sensor, as shown in Figure 4. The vane is mounted to a fixture which is driven by a DC servo motor so that both concave and convex airfoil can be measured. The measurement points on the airfoil surface can be calculated from four-axis coordinates and the sensor output. In order to obtain reliable readings from the laser sensor, the XYZ table and the fixture motor should be controlled in such a manner that the laser beam is normal to the surface. By averaging many measurements at the same point, a very good measurement repeatability of about 5 microns can be observed. Since there is no contact between the sensor and workpiece, the measurement process is continuous, hence very fast. However, the brazed airfoil has very rough surface with shining spots, and the laser sensor does not always produce reliable readings.

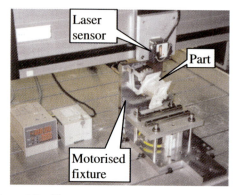

Figure 4 Profile measurement using laser range sensor.

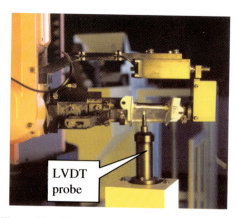

Figure 5 In-situ profile measurement using LVDT.

The alternative approach is to use a contact probe commonly deployed for a Coordinate Measurement Machine (CMM) or a machining centre. Such a measurement system is slow, but the physical contact between the probe and workpiece ensures the reliability of the measurement. A Linear Variable Differential Transducer (LVDT) has been integrated into the robotic system to carry out the profile measurement. The sensor outputs differential square wave signals, 90 degrees phase different. These signals are connected to a linear gauge counter, which outputs Binary-Coded

Decimal (BCD) to a data acquisition board resident in the robot controller. The BCD reading is proportional to the profile sensor displacement. The sensor has a measurement range of 10 mm, resolution of 5 µm, and measurement accuracy of 20 µm.

After gripping the part, the robot approaches the measurement probe, as shown in Figure 5. The sensor measures a number of points in specific cross sections that will be used by the Optimal Profile Fitting (OPF) later. In our development, three sections are selected, as shown in Figure 6.

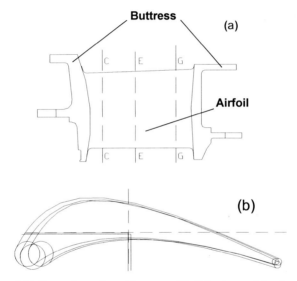

Figure 6 (a) Front view of a turbine vane, (b) Side views of Section C-C, E-E and G-G of the vane airfoil.

For each sectional profile, five measurement points are taken from the concave side and five from the convex side. When the measurement points fall into the brazed area, an approximation is made to offset the measured points to the prior-to-braze airfoil surface. The desired height of the airfoil leading edge is determined by probing the buttress of the workpiece.

To obtain reliable measurement data, the workpiece is positioned to have a normal contact with the sensor probe during each measurement. The sensory displacement readings only give the displacements in the $Z$ axis. In order to obtain the actual coordinates $(X, Y, Z)$ of the measured points, corresponding robot coordinates have to be used for computation in conjunction with displacement readings.

Since the measurement unit is stationary, the profile measurement data is directly given in the global coordinate system, with the Z coordinate being offset by the sensor displacement. For ease of implementation, the profile measurement data have to be transformed from the global coordinate system into the robot hand coordinate system.

## 3.3    *Coordinate Transform*

Assume that the robot global coordinate system is A: *XYZ*, and the gripper (local) coordinate system is B: *X'Y'Z'*. The former is attached to the base of the robot, while the latter is attached to the robot end-effector. The transformations between the global and local coordinate system, as shown in Figure 7, can be readily derived [16].

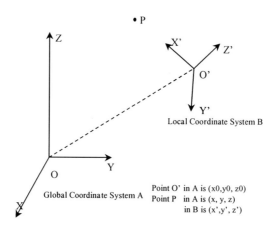

Figure 7 Coordinate system transformation.

Let the coordinates of the point *P* be (*x, y, z*) in the global coordinate system, and (*x', y', z'*) in the local coordinate system. The origin *O'* of the local coordinate system *B* is ($x_0, y_0, z_0$) in the global coordinate system. Then the coordinate transformation from the global coordinate system to the local coordinate system is:

$$\begin{bmatrix} x' \\ y' \\ z' \end{bmatrix} = T \begin{bmatrix} x - x_0 \\ y - y_0 \\ z - z_0 \end{bmatrix} \tag{1}$$

where

$$T = \begin{bmatrix} A_{11} & A_{21} & A_{31} \\ A_{12} & A_{22} & A_{32} \\ A_{13} & A_{23} & A_{33} \end{bmatrix} \tag{2}$$

The matrix T is called rotational transformation matrix representing the rotation from the global coordinate system to the local coordinate system. The matrix elements are computed as follows:

$$\begin{aligned} A_{11} &= \cos(\theta_{xx'}) & A_{21} &= \cos(\theta_{yx'}) & A_{31} &= \cos(\theta_{zx'}) \\ A_{12} &= \cos(\theta_{xy'}) & A_{22} &= \cos(\theta_{yy'}) & A_{32} &= \cos(\theta_{zy'}) \\ A_{13} &= \cos(\theta_{xz'}) & A_{23} &= \cos(\theta_{yz'}) & A_{33} &= \cos(\theta_{zz'}) \end{aligned} \tag{3}$$

where $\theta_{xx'}$ is the angle between axis $X$ and axis $X'$, and the other angles are defined in the similar notions. Based on Equation (1) to (3), together with the $Z$-axis offset given by the sensory reading, the coordinates of a measurement point can be transformed from the global coordinate system to the hand coordinate system.

## 4. Template-Based Optimal Profile Fitting

### 4.1 *Template Generation*

The design profiles of the workpiece are given in the form of 2D coordinates of 35 points for each of left, middle and right cross sections (Figure 6). To obtain a complete sectional profile we apply the cubic spline interpolation to the given design points.

Cubic spline interpolation ensures that not only the interpolation is continuously differentiable on the interval, but also that it has a continuous second derivative on the interval.

Given a function $f$ defined on the nodes $x_i$ ($i=0,1,\ldots,n$), a cubic spline interpolation $S$ is a function that satisfies the following conditions:

1. $S$ is a cubic polynomial, denoted by $S_i$, on the subinterval $[x_i, x_{i+1}]$.
2. $S(x_i) = f(x_i)$   $i = 0,1,\cdots,n-1$.
3. $S_{i+1}(x_{i+1}) = S_i(x_{i+1})$   $i = 0,1,\cdots,n-2$.

4.  $S'_{i+1}(x_{i+1}) = S'_i(x_{i+1})$  $i = 0,1,\cdots,n-2$.

5.  $S''_{i+1}(x_{i+1}) = S''_i(x_{i+1})$  $i = 0,1,\cdots,n-2$.

6.  $S'(x_0) = f'(x_0)$, and $S'(x_n) = f'(x_n)$ (clamped boundary).

To construct the cubic spline interpolation for a given function $f$, the conditions above are applied to the cubic polynomials:

$$S_i(x) = a_i + b_i(x - x_i) + c_i(x - x_i)^2 + d_i(x - x_i)^3$$

For donation convenience, let $h_i = x_{i+1} - x_i$. Then,

$$S_i(x_i) = a_i = f(x_i)$$
$$S_{i+1}(x_{i+1}) = S_i(x_{i+1}) = a_{i+1} = a_i + b_i h_i + c_i h_i^2 + d_i h_i^3$$

hold for $i=0,1,\ldots,n-2$.  Applying the condition 4, we have

$$b_{i+1} = b_i + 2c_i h_i + 3d_i h_i^2 \qquad\qquad i = 0,1,\cdots,n-1$$

Define $c_n = {S''(x_n)}\big/{2}$ and apply the condition 5, then

$$c_{i+1} = c_i + 3d_i h_i \quad i = 0,1,\cdots,n-1$$

Solving for $d_i$ in the above equations gives

$$a_{i+1} = a_i + b_i h_i + \frac{h_i^2}{3}(2c_i + c_{i+1})$$
$$b_{i+1} = b_i + h_i(c_i + c_{i+1})$$

Finally, re-arranging the equation gives

$$h_{i-1}c_{i-1} + 2(h_{i-1} + h_i)c_i + h_i c_{i+1} = \frac{3}{h_i}(a_{i+1} - a_i) - \frac{3}{h_{i-1}}(a_i - a_{i-1})$$

In matrix form, the above equation becomes

$$
\begin{bmatrix}
2h_0 & h_0 & 0 & 0 & \cdots & 0 \\
h_0 & 2(h_0 + h_1) & h_1 & 0 & \cdots & 0 \\
0 & h_1 & 2(h_1 + h_2) & h_2 & \cdots & 0 \\
\vdots & \ddots & \ddots & \ddots & \ddots & \vdots \\
0 & \cdots & 0 & h_{n-2} & 2(h_{n-2} + h_{n-1}) & h_{n-1} \\
0 & 0 & \cdots & & h_{n-1} & 2h_{n-1}
\end{bmatrix}
\times
\begin{bmatrix}
c_0 \\ c_1 \\ c_2 \\ \vdots \\ c_{n-1} \\ c_n
\end{bmatrix}
$$

$$
=
\begin{bmatrix}
\dfrac{3}{h_0}(a_1 - a_0) - 3f'(x_0) \\[2mm]
\dfrac{3}{h_1}(a_2 - a_1) - \dfrac{3}{h_0}(a_1 - a_0) \\[2mm]
\vdots \\[1mm]
\vdots \\[2mm]
\dfrac{3}{h_{n-1}}(a_n - a_{n-1}) - \dfrac{3}{h_{n-2}}(a_{n-1} - a_{n-2}) \\[2mm]
3f'(x_n) - \dfrac{3}{h_{n-1}}(a_n - a_{n-1})
\end{bmatrix}
$$

where the clamped-boundary condition (6) has been used.

The above linear equation has a unique solution for $c_0, c_1, \ldots, c_n$ as the linear system coefficients matrix is strictly diagonally dominant. Through the values of $a_i$ and $c_i$ ($i = 0, 1, \ldots, n$), the values of $b_i$, $d_i$, and $S_i$ ($i = 0, 1, \ldots, n\text{-}1$) can be easily computed. The 2D coordinates of any point on the design profiles in robot coordinate system can then be calculated.

Using the above Cubic Spline Interpolation, the approximate profiles of the three sections of the master workpiece can be plotted, as illustrated in Figure 6. These profiles are used as templates for generation of actual profiles in a distorted workpiece.

## 4.2 Profile Fitting Requirements

Since the workpiece is gripped by the robot during the profile measurement and blending operations, the optimal profile fitting is performed in the robot hand coordinate system. In the Data-Driven Supervisory Controller (DDSC), the Optimal Profile Fitting (OPF) module fits a template to the actual measurement points with minimum sum of errors. The sectional template profiles are established based on design data using Cubic Spline Interpolation. The Cubic Spline Interpolation has a well-defined mathematical structure.

The 3D airfoil surface fitting can be performed in two stages. In the first stage, a number of specified 2D sectional airfoil profiles are fitted individually. A multi-step template-based optimal profile fitting method is used in each sectional profile fitting. In the second stage, a 3D airfoil profile is generated from interpolating the 2D sectional profiles. The computations involved in the second stage are relatively easy and the details are therefore omitted in this chapter.

Due to the complicated nature of the problem, it is required that the airfoil sectional profile fitting method must have the following capabilities.

1.  The fitting method should be able to fit a sectional airfoil profile with distortions not only in the aspects of the position and orientation, but also in the shape of the profile.
2.  A complete smooth sectional airfoil profiles should be generated, despite the fact that only certain portion of the profile can be measured and the number of the measurement points is limited.

During the profile fitting, each measurement point should be treated differently according to its location. This requirement is due to the fact that most of the airfoil surface is covered by the brazing material to be ground/polished away. There are large variations in both the thickness and the position of the brazing material. The measurement data from those covered areas are more noisy, while those for the uncovered area are much more accurate.

For some delicate portions of the airfoil, such as the trailing edge of the turbine vane, the profile must be fitted with a high precision. On the other hand, the repair areas of the airfoil should be loosely fitted. Contrary to the objective of fitting, certain corrections from the distortions are made so that the shape of the fitted profile has less variation from that of the design profile. By examining the above stringent requirements, it is clear that normal regression or interpolation methods cannot be applied for this profile fitting problem. A different fitting strategy must be designed.

In search of a new fitting method, we notice the fact that, although the actual airfoil profiles of the turbine vanes are largely distorted, they still retain, at least locally, the major features of their design profiles. Otherwise, they would not be repairable and have to be scrapped. It is desirable to extract the useful information from the design profile and utilise it in the generation of the distorted profile. Based on this observation, we developed a new template-based optimal profile fitting method.

The optimal profile fitting is performed in multiple steps. In each step, the designed sectional airfoil profile is used as a rigid 2D template, which is a smooth airfoil profile generated from its engineering data by cubic spline interpolation. The template is shifted and rotated in the 2D plane to the optimal position such that a weighted sum distance from the measurement points to the template is minimum. There are three degrees of freedom in moving the template, i.e. (i) *X*-axis shift, (ii) *Y*-axis shift, and (iii) rotation. As such, the template-based optimal profile fitting in each step becomes a multi-dimensional minimisation process. A new fast converging direct search minimisation algorithm, which will be discussed in detail in the next section, is used for the optimal profile fitting.

The index function for the minimisation is the weighted sum distance given by

$$d_{sum} = w_h \, h + \sum_{i=1}^{n} w_i \, d_i \qquad (4)$$

where $d_{sum}$ is the weighted sum distance, $n$ is the total number of the measurement points on convex and concave sides of the airfoil profile, $d_i$ is the distance from *i*th measurement point to the template, $h$ is the height difference between the measured leading edge and that of the template, $w_i$ and $w_h$ are the weightages for the measurement points and the leading edge height, respectively. Notice that the height of the airfoil leading edge is treated separately in Equation (4). This is because the airfoil leading edge cannot be measured directly and its height can only be derived from measurements on the buttress of the workpiece.

Different weightages are properly chosen for different measurement points. In this way, the measurement points for different portions of the profile, which are of different accuracies, can be treated differently. By the same means, some delicate portions of the profile can be fitted with a high precision. While for other portions, the profile is loosely fitted, so that those portions of the fitted profile have less shape distortions as compared with the design profile.

To overcome the distortions in the shape of the profile, a series of template-based optimal profile fittings are performed. Each step of the profile fitting is targeted at a different portion of the profile. By varying the weightages $w_i$ and $w_h$ for different fitting steps, the template is fitted to different portions of the profile accordingly. When all the portions of the profile are fitted, the fitting results are appropriately combined together to form a smooth complete profile. The final fitted profile satisfies both the contradictory objectives of profile fitting and distortion correction. Figure 8 shows the schematics of actual 2D profile obtained by OPF. The design data are used as the template, and the sectional profile is derived from the measurement points.

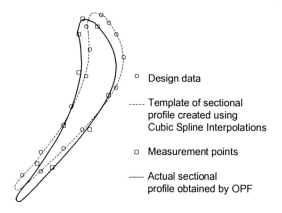

Figure 8 Actual 2D sectional profile obtained by OPF.

## 4.3    *A Fast Converging Minimisation Algorithm*

As mentioned in the last section, the optimal profile fitting of a 2D sectional airfoil profiles needs to do unconstrained minimisation in a 3D search space with $X$-axis shift, $Y$-axis shift and rotation as the three independent variables. In this section, we present a generic fast converging multi-dimensional minimisation algorithm.

There are a number of multi-dimensional minimisation algorithms, such as downhill simplex search method, Hooke-Jeeves pattern search method, Powell conjugate direction method, Cauchy method, Newton method, conjugate gradient methods, and variable metric methods [10, 11]. However, for our particular minimisation problem the selection of the methods is limited due to the complication of the index function.

The index function in the optimal profile fitting, as given by Equation (4), is the weighted sum distance from the measurement points to the template. The template is represented by a series of points. The three position parameters of the template, i.e., $X$-axis shift, $Y$-axis shift and rotation, are independent variables of the index function. As the airfoil profiles are odd shaped, there is no analytical expression available for the index function. As a matter of fact, the value of the index function can only be obtained through some numerical methods. Therefore, any existing minimisation method which uses analytical expression or any derivatives of the index function will not be applicable to our problem.

To solve airfoil optimal profile fitting problem, we developed an intuitive direct search method. It turns out that the new method is an improved version of Hooke-Jeeves pattern search method [12], with a faster convergence rate. The basic idea is that we take a point in the search space as a base point and explore around it to find a right direction and a right step to move the base point towards the minimum point. In order to obtain a fast convergence rate, the direction is adjusted and the step is increased repeatedly before each move of the base point. The following is the outline of the multi-dimensional minimisation algorithm. An illustrative flow chart of the algorithm is shown in Figure 9.

Step 1   Set initial base point and initial search span. The search span may be different for each coordinate direction in the search space.

Step 2   Test the index function values of the surrounding points of base point with search span. Find the best point (i.e. with the lowest index) among the surrounding points. In this step, all the coordinate directions in the search space have to be exhausted.

Step 3   Compare *base point* with *best point*. If *base point* is better (i.e. with a smaller index) than *best point* (which means the minimum point is within the search span), jump to Step 9. Otherwise, continue.

Step 4   Take *best point* as *direction point*.

Step 5   Along the direction from *base point* to *direction point*, compute *temporary base point* by extrapolation with a ratio, say 2.5. That means *temporary base point* will be away from *base point* 2.5 times of the distance between *base point* and *direction point*.

Step 6   Test the index function values of the surrounding points of *temporary base point* with *search span*. Find *best point* among the surrounding points. Again, all the coordinate directions in the search space have to be considered in this step.

Step 7   Compare *direction point* with *best point*. If *best point* is better than *direction point* (which means that the direction of last extrapolation is probably right and it is worth to further increase the move step of *base point*), jump to Step 4.  Otherwise, continue.

Step 8   Take *direction point* as *base point* and jump to Step 2.

Step 9   Reduce *search span*, say, 10 times.

Step 10 If *search span* is not sufficiently small, jump to Step 2. Otherwise, continue.

Step 11 Take *base point* as the minimum point. End of the program.

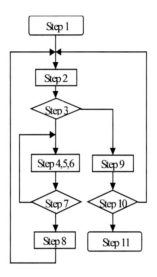

Figure 9 Flow chart of direct search minimisation algorithm.

In the above minimisation algorithm, the loop involving Steps 2-8 is for repeated move of the base point towards the minimum point.  The inner loop involving Steps 4-7 is for repeated increase of the moving step for the base point.   The loop involving Steps 2, 3, 9 and 10 is for repeated reduction of the search span around the base point.

Notice that with the inner loop involving Steps 4-7, the moving step of the base point towards the minimum point increases very fast.  If the moving direction remains the same in the iterations, the step increases faster than an exponential function and is given by:

$$l = \lambda \sum_{i=0}^{n} a^i \qquad\qquad (5)$$

where $l$ is the moving step of the base point, $\lambda$ is the distance from base point to direction point at the first iteration (the search span at the moving direction), $a$ is the extrapolation ratio selected in Steps 5 and $n$ is the iteration number of the loop involving Steps 4-7.

With this fast increase of moving step of the base point, a large reduction rate for the search span, say 10 times, can be used in Step 10. With the features of fast increase of the moving step for base point and large reduction rate for search span, the above algorithm gives a faster convergence rate than that proposed by Hooke-Jeeves [12].

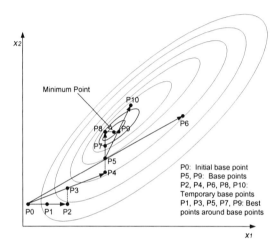

Figure 10 Direct search minimisation in 2-D space.

Figure 10 shows the minimisation process of the algorithm for a two-dimensional case with the independent variables $x_1$ and $x_2$. Notice the fast movement of the base point toward the minimum point starting from $P_0$ to $P_5$ and then to $P_9$. As $P_9$ is "better" than all its surrounding points, the search span is reduced 10 times. Then, taking $P_9$ as a new base point, the minimisation process will continue with the reduced search span. The minimisation process ends when the search span is sufficiently small.

## 4.4   *Software Development*

The template-based optimal profile fitting method is implemented using object-oriented program techniques. The program is coded in a class called

CFitting and compiled using Microsoft Visual C++ with Microsoft Foundation Classes (MFC).  Together with other program modules, the algorithm is integrated into the main program AutoBlending.exe running on the host computer.

In the implementation of the main program, C++ classes are used in combination with C++ call function modules.  The following classes are implemented for those algorithms with specific tasks and those that need to manipulate some data for particular purposes.

**CautoBlendingDlg** – This MFC derived class is the main class of the program and uses various global functions and class member functions to perform all kinds of tasks. The associated dialog serves as graphic user interface (GUI) for the final system.

**CdigitalIO** – It is a class dealing with all the functions related to digital input and output between the robot controller and the host computer.

**Cprofile** – It uses cubic spline interpolation method and generates the smooth design airfoil profiles (the templates for the profile optimal fitting) from the given engineering data.

**Cfitting** – It implements the template-based optimal profile fitting algorithm.

Computations related to robotic coordinate transformations and other computations, such as file operations and system calibrations, are implemented in the call functions.  Details are omitted in this chapter.

In the airfoil profile fitting, two steps of the template-based optimal profile fitting are performed.  One step is targeted at the convex side of the airfoil, and the other is targeted at the concave side.  The weightages of measurement points in the sum distance in Equation (4) for the two optimal fitting steps are as shown in Table 1.

Table 1 Weightages used in sum distance in multi-step template-based optimal profile fitting.

|  | $w_1$ | $w_2$ | $w_3$ | $w_4$ | $w_5$ | $w_6$ | $w_7$ | $w_8$ | $w_9$ | $w_{10}$ | $w_h$ |
|---|---|---|---|---|---|---|---|---|---|---|---|
| Concave fitting | 10 | 2 | 2 | 2 | 2 | 1 | 1 | 1 | 1 | 1 | 10 |
| Convex fitting | 1 | 1 | 1 | 1 | 1 | 2 | 2 | 2 | 2 | 10 | 10 |

Notice that measurement points No. 1 to No. 5 are on the concave side of the airfoil, and points No. 6 to No. 10 are on the convex side, as shown in Figure 11. Measurement points No. 1 and No. 10 are on the trailing edge, where there is no brazing material on the surface and accurate profile fitting is required. Therefore, large weightages are used for measurement points No. 1 and No. 10. Large weightages are also used for the heights of the airfoil leading edges, as the leading edges need to be fitted with a high precision and the measurement data for the heights are accurate. Measurement points No. 2 to No. 9 are all on the surface with the brazing material, and corrections of the shape distortions for those portions of the profile is required. Therefore, small weightages are used for measurement points No. 2 to No. 9.

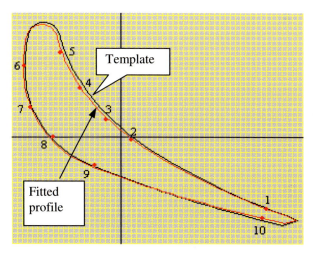

Figure 11 A typical result of multi-step template-based optimal profile fitting.

With proper selection of the weightages, the stringent and contradictory objectives of accurate profile fitting for the airfoil trailing edges and distortion correction for other portions of the profile can be simultaneously satisfied. In other words, the final fitting result is an optimal trade-off between the two contradictory objectives. A typical fitting result is shown in Figure 11. It shows that a profile is tightly fitted to the trailing edge (point #1 and #10), but loosely fitted in the brazed area (point #2 to #9).

## 5.    **Adaptive Robot Path Planner**

### 5.1    *Definition of Tool Path*

Having an accurate description of the airfoil profile is not the ultimate aim. The computed profile based on the sensory data must be used to automatically generate the robotic blending path, which can be understood and accepted by the robot controller. Position control is required for holding the position of a vane in the feed direction against the tangential cutting force and maintaining adequate in-feed in the normal direction. This can minimise errors caused by part distortions and braze layer variations.

In a conventional CNC system, an NC program based on a CAD model can accurately drive the rigid NC machine. However such control methodology cannot be effectively applied for a 6-axis robot for blending individual parts. Thus in order to mimic the capability of a human polisher, the blending path has to be adaptively planned according to individual part conditions.

The space curve that the robot end-effector moves along from the initial location (position and orientation) to the final location is called the *path* [17]. It deals with the formalism of describing the desired robot end-effector motion as sequences of points in space (position and orientation of the robot end-effector) through which the robot end-effector must pass, as well as the space curve that it traverses.

The desired tool path is described by a series of points (the endpoints and intermediate points) in Cartesian coordinates $(X, Y, Z, A, B, C)$, as shown in Figure 12. Such a description is chosen because visualising the correct configurations in Cartesian coordinates is much easier than in joint coordinates.

Along a curve between any two points, the robot automatically moves using the cubic spline line motion. Thus, we are only concerned about the formalism of deriving the robot coordinates $(X, Y, Z, A, B, C)$ of the path points which the robot must travel along in the Cartesian coordinate system. The coordinates $(X, Y, Z)$ specify the robot end-effector position and coordinates $(A, B, C)$ specify the robot end-effector orientation. The coordinates of every path point are calculated based on the robot kinematics model. In addition, certain system set-up parameters, such as polishing wheel size and global locations of all PCT heads, have been built into the model, and can be rapidly and accurately re-configured through Intuitive Tool Calibration (ITC).

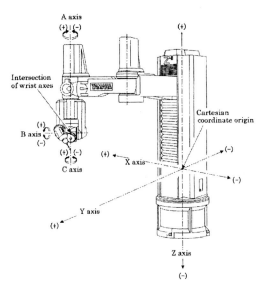

Figure 12 Path point specifications in Cartesian coordinates.

## 5.2 *Derivation of End-Effector Orientation*

A $3 \times 3$ rotation matrix can be defined as a transformation matrix to describe and represent the rotational operations of the robot end-effector's coordinate system with respect to the global coordinate system which is located at the base of the robot. The rotation matrix simplifies many computational operations, but it needs nine angles to completely describe the orientation of a robot end-effector. Instead we employ three Euler angles [18] *A*, *B* and *C* to describe the orientation of the robot end-effector with respect to the global coordinate system. There are many different types of Euler angle representations. The scheme chosen in this development is shown in Figure 13.

The sequence of rotations for the Euler angles is very important.

(1) A rotation of *A* angle about the *OZ* axis:

$$R_{Z,A} = \begin{bmatrix} \cos A & -\sin A & 0 \\ \sin A & \cos A & 0 \\ 0 & 0 & 1 \end{bmatrix}$$

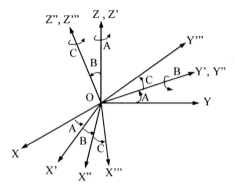

Figure 13 Euler angles system.

(2) A rotation of $B$ angle about the rotated OY' axis:

$$R_{Y',B} = \begin{bmatrix} \cos B & 0 & \sin B \\ 0 & 1 & 0 \\ -\sin B & 0 & \cos B \end{bmatrix}$$

(3) Finally a rotation of $C$ angle about the rotated $OZ''$ axis:

$$R_{Z'',C} = \begin{bmatrix} \cos C & -\sin C & 0 \\ \sin C & \cos C & 0 \\ 0 & 0 & 1 \end{bmatrix}$$

The resultant Euler rotation matrix is obtained by multiplying the above three rotation matrices together:

$$R_{A,B,C} = R_{Z,A} R_{Y',B} R_{Z'',C}$$
$$= \begin{bmatrix} \cos A \cos B \cos C - \sin A \sin C & -\cos A \cos B \sin C - \sin A \cos C & \cos A \sin B \\ \sin A \cos B \cos C + \cos A \sin C & -\sin A \cos B \sin C + \cos A \cos C & \sin A \sin B \\ -\sin B \cos C & \sin B \sin C & \cos B \end{bmatrix}$$

With the above defined Euler angle rotation matrix, the orientation of the robot end-effector can be derived with respect to the reference global coordinate system.

## 5.3    *Generation of Tool Path*

Combining with the translation of the robot end-effector, the tool path point coordinates ($X$, $Y$, $Z$, $A$, $B$, $C$) in global coordinate system can be derived through coordinate system transformations. An illustration of robot path generation is given in Figure 14.

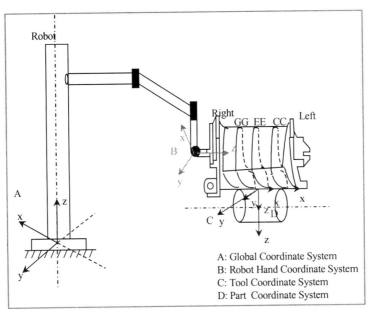

Figure 14 Part coordinate system.

There are four 3D coordinate systems:

1.  Coordinate system A: Robot global coordinate system CS.
2.  Coordinate system B: Robot hand (end-effector) CS.
3.  Coordinate system C: Tool CS.
4.  Coordinate system D: Part CS, constructed for each blending position as shown in Figure 15.

After the part's 3D profile is obtained, using the Optimal Template Fitting method, the part coordinates system is constructed on the 3D profile for each blending point. The part and tool coordinate systems are constructed such that, for each blending point, the part is at desired blending position when the two coordinate systems are overlapped.

Applying this constraint, we can derive the position of robot end-effector's coordinate system B and furthermore the robot coordinates (X, Y, Z, A, B, C) by coordinate system transformations.

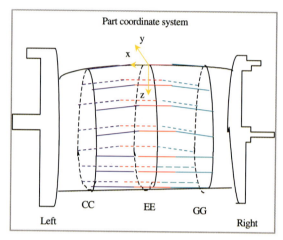

Figure 15 Robot path generation.

The robot path generation is the inverse coordinate transformation obtained by translation and Euler angles rotation. The following is the procedure to compute the robot coordinates of a path point:

1. Compute the path points in robot end-effector coordinate system.
2. Construct a part coordinate system for a path point.
3. Assuming that the part coordinate system is overlapped with the tool coordinate system, compute the robot end-effector coordinate system using the coordinate system transformation.
4. Derive the robot coordinates (X, Y, Z, A, B, C) from the position of robot coordinate system in the global coordinate system.
5. Repeat Step 2 to Step 4.

Through the above steps, a series of robot coordinates (X, Y, Z, A, B, C) are obtained for all desired blending positions, which forms the robot blending path. The robot path generation algorithm is implemented in the C++ class **CPathGeneration**. Figure 16 shows one tool path point computed by ARP. Before the robot path is downloaded to the robot controller, computer simulation can be carried out to verify individual path points.

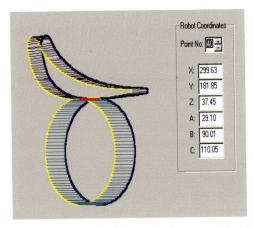

Figure 16 Computer simulation of path generated by ARP.

## 6. Implementation of SMART 3D Blending System

After the successful technology development, the research team implemented the first prototype system - a **S**elf-compliance, **M**ulti-tasking, **A**daptive-planning, **R**econfigurable and **T**eaching-free (**SMART**) robotic system for polishing distorted 3D turbine airfoils. The overview of the system is shown in Figure 17.

Figure 17 Overview of the SMART 3D Blending System.

Corresponding to the conceptual design described in Chapter 2, the system consists of an index table, an In-Situ Profile Measurement (IPM) station using a LVDT, a Yamaha Z-II6 finishing robot, and four Passive Compliant Tools (PCT). Various control modules implemented empower the system with advanced automation capabilities.

## Self-Compliance

The Self-Aligned End-Effector (SAE) enables the robot to grip the workpiece in a fixed axial alignment, which reduces gripping errors. The SAE is driven by the end axis of the robot, and provides a full rotation of 360 degrees. Built-in sensors can detect any misalignment and poor gripping. The system employs Passive-Compliant Tooling (PCT) to control the contact force between the SAE and the polishing head. This capability imitates human muscular control of polishing force to avoid any over-cutting or under-cutting.

## Multi-Tasking

The measurement data have to be transferred from the robot controller to the host controller for numerical computation. The computed polishing path data are then downloaded to the robot controller for executing polishing tasks. A sequential execution of the communication and polishing tasks would stretch the cycle time. To eliminate the waiting time and shorten the production cycle time as required, a multi-tasking robot control system has been developed, which allows the communication tasks to be executed in parallel with the polishing tasks.

## Adaptive Planning

The SMART robotic system is built with adaptive compensation capability analogous to human's visual adaptation to part distortions and variations, through implementation of the In-Situ Profile Measurement (IPM), the template-based Optimal Profile Fitting (OPF) algorithm, and the Adaptive Robot Path Planner (ARP). Accurate process models have been implemented in the Knowledge-Based Process Controller and compensate errors arising from variable tooling conditions and process dynamics.

## *Reconfigurable*

The SMART system can be easily configured to polish different kinds of turbine vane products. The product configuration data can be stored in the data base. By simply pressing buttons on a touch screen, a particular airfoil class and brazing pattern can be chosen for the current production from up to 210 combinations of the airfoil classes and brazing patterns. The SMART system has the flexibility to reconfigure the tooling and measurement systems. Different polishing tool heads and profile measurement sensors can be deployed. Any drifts in the positions of the sensor and tool heads in the global reference coordinate system can be manually keyed in through Intuitive Tool Calibration (ITC). Hence sensor data transformation and robot kinematics can be readily computed based on the calibration readings during reconfiguration.

## *Teaching-Free*

Despite part geometric uncertainties, the SMART system can automatically accommodate the variations due to its capabilities of adaptive planning and knowledge inference. Time and labour-consuming manual teaching is no longer required. This leads to ease of use, better utilisation and higher productivity. The automated system shortens the cycle time from 10 minutes (manual polishing) to an average of 5.75 minutes, resulting in an improvement of 42.5%.

## 7.    Results

Before being deployed for production use in the shop floor, the SMART system had gone through rigorous qualification tests against all quality requirements. It has been benchmarked for the first stage vane of JT9D-7Q engine. Three categories of quality measures have been examined, i.e. dimensions of the finish profile, the surface roughness and finish quality, and the wall thickness.

## 7.1    *Dimension of Finish Profile*

Actual cross sectional profiles were derived by fitting the templates (design profiles) into the measurement points, and compared with the design profiles. As show in Figure 18, the difference between the templates and corresponding measurement points can be as large as 1 mm. Then the test pieces were measured by the In-Situ Profile Measurement system before blending and after blending.

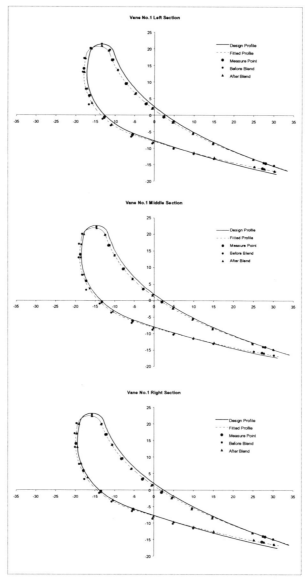

Figure 18 Sectional profiles derived from data, measurement data prior to blending, and measurement data after blending.

Despite the distortions and shifting of the cross sectional profiles, the final finish profile (measured after blending) follows the measured points closely. Material removal at the trailing edge is kept to a minimum to avoid any overcutting into the thin and sensitive area, while smooth transition from the non-brazed area to the brazed area in guaranteed, as evidently shown in Figure 18. All measurement points and those before blending and after blending fall on the fitted profiles.

Figure 19 shows the cutting depth at the measurement points. At measurement point #20 (around the leading edge), the cutting depth is as high as 1 mm, while at points #1, #2, #3 (around the trailing edge), the cutting depth is less than 0.05 mm. The gap between the leading edge and the template (quantifying the leading edge position) is within ±0.20 mm that is significantly smaller than 0.50 mm specified. The feature of such adaptive cutting is the key to achieving the desired finish profile without overcutting and undercutting.

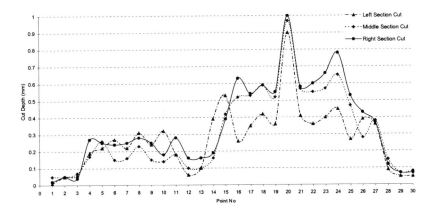

Figure 19 Cutting depth at measurement points.

## 7.2    *Surface Roughness and Finish Quality*

The surface roughness of six vanes were measured on both concave and convex sides, and tabulated in Table 2. The average roughness ranges from 1.022 to 1.4 microns $R_a$, better than the required 1.6 microns $R_a$.

Vanes before and after blending are shown in Figure 20. A very smooth airfoil profile was achieved by the SMART system. Further visual inspection shows no visible transition lines from the non-brazed area to the brazed one, no visible blending marks in the cutting path overlap areas, and no burning marks. The curvature transition from the concave to convex airfoil is very smooth and more consistent than the one generated by manual blending.

Table 2 Roughness measurement results.

| Vane no. | $R_a$ (μm) | | Average $R_a$ (μm) |
|---|---|---|---|
| | Convex | Concave | |
| 1 | 1.378 | 1.417 | 1.400 |
| 2 | 1.059 | 1.090 | 1.075 |
| 3 | 1.228 | 1.086 | 1.157 |
| 4 | 1.003 | 1.041 | 1.022 |
| 5 | 1.197 | 0.936 | 1.067 |
| 6 | 1.196 | 0.980 | 1.088 |

Figure 20 Vanes before (left) and after (right) robotic grinding and polishing.

It is worth noting that the consistency of finish profiles also benefits the downstream laser drilling operation. If the finish profile is not consistent in terms of wall thickness, airfoil shape and leading edge position, the cooling holes generated by the laser machine may converge, and affect the aerodynamic performance of the engine, and in many cases be rejected by Quality Assurance. In the automated system, such failure is minimised, if not eliminated.

## 7.3    *Wall Thickness*

Following the existing practice in the aerospace overhaul industry, the wall thickness needs to be quantified. Minimum wall thickness must be maintained in order not to undermine the airfoil strength. In the selected case, seven points are measured against the minimum wall thickness, points #1 to #7 in Figure 21.

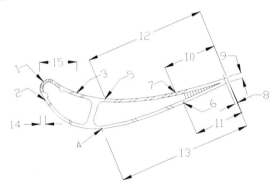

Figure 21 Positions of checking points.

Figure 22 plots the minimum wall thickness, and the measurement results for three vanes. It shows that all measurements at the designated seven points are above the minimum wall thickness. Figure 23 shows the samples of sectioned airfoils. Trailing edge thickness is maintained, and the smooth curvature of leading edge is confirmed.

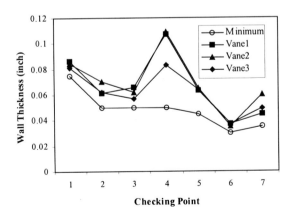

Figure 22 Results of wall thickness measurement.

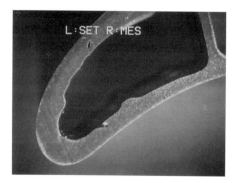

Figure 23 Cross sections of polished vanes: leading edge (left) and trailing edge (right).

## 8.     Concluding Remarks

In a concerted effort, we have successfully developed a Knowledge-Based Adaptive Robotic System for 3D Profile Grinding and Polishing. Around the finishing robot, Self-Aligned End-effector (SAE), Passive Compliant Tools (PCT) and In-Situ Profile Measurement (IPM) system have been developed. Template-based Optimal Profile Fitting (OPF) algorithm, and Adaptive Robot Path Planner (ARP) have been developed to overcome the part distortions inherent in component overhaul and to satisfy force and compliance control required for robotic blending. The technological modules have been built into the first working prototype SMART 3D Grinding/Polishing System.

The synergistic combination of hardware and software solutions enables the SMART system to meet stringent quality requirements and design criteria, such as profile smoothness, surface roughness, leading edge transition and height, minimum wall thickness, removal of transition lines between brazed and non-brazed areas, etc. The SMART system, the first of its kind for blending distorted airfoils, has been benchmarked against critical quality measures, and satisfies all product and process requirements. It shortens the cycle time from 10 minutes (manual blending) to an average of 5.75 minutes, resulting in an improvement of 42.5%. It has since been installed for production use.

The technological breakthrough lays a strong foundation for future explorations of robotic machining for other advanced applications, such as overhauling fan blades (currently done by operator-assisted CNC), and manufacturing new aircraft components. Another challenge is to automate the finishing process of precision mechanical components such as 3D moulds, which is still accomplished manually despite earlier attempts by various schools worldwide.

# References

1.  Engel, T. and Tomastik, R. "Description of the chamfering and deburring end-of-arm tool (CADET)," The Fourth International Conference on Control, Automation, Robotics and Vision, Singapore, 2-6 December 1996, pp. 629 – 633

2.  Berger, U., Janssen, R. and Brinksmeier, E. "Advanced mechatronic system for turbine blade manufacturing and repair", International Conference on Computer Integrated Manufacturing, Singapore, 21-24 October 1997, pp. 1395-1404.

3.  Chen Y.H., and Hu, Y.N. "Implementation of a robot system for sculptured surface cutting. Part 1: rough machining", International Journal of Advanced Manufacturing Technology (1999) 15, pp. 624-629.

4.  Hu, Y.N., and Chen, Y.H. "Implementation of a robot system for sculptured surface cutting. Part 2: finish machining", International Journal of Advanced Manufacturing Technology (1999) 15, pp. 630-639.

5.  Ozaki, F., Jinno, M., Yoshimi, T., Tatsuno, K., Takahashi, M., Kanda, M., Tamada, Y., and Nagataki, S. "A force controlled finishing robot system with a task-directed robot language," Journal of Robotics and Mechatronics, 1995, Vol. 7, No. 5, 1995, pp. 383-388.

6.  Kunieda, M. and Nakagawa, T. "Robot-polishing of curved surface with magneto-pressed tool and magnetic force sensor", Proceedings of 25th International MTDR Conference, April 1985, pp. 193-200.

7.  Woon, L.C., Ge, S.S., Chen, X.Q., and Zhang, C. "Adaptive neural network control of coordinated manipulators", Journal of Robot Systems, April 1999, Vol. 16, No. 4, pp. 195-211.

8.  Lancaster, P., and Salkauskas, K. "Curve and Surface Fitting - An Introduction," Academic Press Inc. (London) Ltd., London, 1986.

9.  Chapra, S.C., and Canale, R.P. "Numerical Methods for Engineers – With Programming and Software Applications," Third Edition, WCB/McGraw-Hill, Boston, 1998.

10. Press, W.H., Flannery, B.P., Teukolsky, S.A., and Vetterling, W.T. "Numerical Recipes in C," Cambridge University Press, New York, 1988.

11. Reklaitis, G.V., Ravindran, A., and Ragsdell, K.M. "Engineering Optimization: Method and Applications," John Wiley & Sons, Inc., New York, 1983.

12. Hooke, R. and Jeeves, T.A. "Direct search of numerical and statistic problems," J. ACM, 1996, Vol. 8, pp. 212-229.

13. Gong, Z.M., Chen, X.Q. and Huang, H. "Optimal profile generation in distorted surface finishing", IEEE International Conference on Robotics and Automation, San Francisco, USA, 22-24 April 2000, pp. 1557-1562.

14. Burden, R.L. and Faires, J.D. "Numerical Analysis," Fifth edition, PWS Publishing Company, Boston, 1993.

15. Pizer, S.M. "Numerical Computing and Mathematical Analysis," Science Research Associates, Chicago, 1975.

16. Gellert, W. and Kustner, H. "The UNR Concise Encyclopedia of Mathematics," Second edition, Van Nostrand Reinhold Company, New York, 1989.

17. Fu, K.S., Gonzalez, R.C. and Lee, C.S.G. "Robotics: Control, Sensing, Vision, and Intelligence," McGraw-Hill Book Company, New York, 1987.

18. Craig, J.J. "Introduction to Robotics: Mechanics and Control," Second edition, Addison-Wesley Publisher Company, Mass, 1989.

19. Chen, X.Q., Gong, Z.M., Huang, H., Ge, S.S. and Zhu, Q. "Development of SMART 3D polishing system" Industrial Automation Journal, A Singapore industrial automation association publication, April-June 1999, pp. 6-11.

20. Yamaha Finishing Robot – User's Manual Version 1.4.

# CHAPTER 4

# ACOUSTIC EMISSION SENSING AND SIGNAL PROCESSING FOR MACHINING MONITORING AND CONTROL

Hao Zeng and XiaoQi Chen

*Gintic Institute of Manufacturing Technology,*
*71 Nanyang Drive, Singapore 638075*

## 1.    Introduction

The manufacturing industry is being driven by demands for increasing product quality, smaller batch sizes, shorter product delivery times, and globalised manufacturing. At the same time, the machining process must run reliably and without disturbances, which is possible if there is an optimal degree of automation and flexibility.

Machining processes are required in various cutting operations, either rough cutting or precision finishing of a workpiece. These processes have to deal with deformation of the cutting tool, the machine tool, and the workpiece, which is caused by cutting force, thermal effect and chatter vibration. Furthermore, tool wear and tool failure are prominent in machining difficult-to-cut materials such as Inconel and Titanium [1]. The optimum performance of these complex processes relies on the availability of the data about process conditions for process monitoring and feedback to the process controller. It further demands reliable sensing systems coupled with robust signal processing techniques to extract useful information from a machining process. This is especially true in consideration of a precision machining operation where very stringent tolerances are required. Under such a dynamic cutting condition, it is very difficult to predict, hence control in an open loop, the machine behaviour. Therefore a process monitoring system is required to identify the actual cutting conditions through in-process sensors. Based on the sensory information, adaptive machining control, ranging from simple feed stop to advanced process control, can be implemented. An in-process monitoring and control system offers a number of benefits:

- Improve the finishing accuracy and consistency, consequently reducing reworks or rejects.
- Enhance the machining efficiency and reduce the machine idle time.
- Increase the tool life.
- Prevent damage to the machine.

The subjects to be monitored or controlled can be categorised into four categories: machine, tool, process, and workpiece. In each of these categories, the machining conditions are divided into real-time and non real-time, as summarised in Table 1. The real-time operations require a system response time within a range of milliseconds while the non real-time operations may take seconds or even minutes.

Table 1 Monitoring conditions in a machining process.

|  | Non real-time monitoring | Real-time monitoring |
|---|---|---|
| **Machine** | Accuracy<br>Thermal deformation | CNC motion control,<br>Collision detection |
| **Process** | Coolant<br>Chip formation<br>Temperature | Chatter/Vibration, power,<br>cutting force/torque, cutting<br>path /parameters |
| **Tool** | Tool wear measurement and<br>compensation (offset) | Tool fracture, tool wear status<br>for optimal process control |
| **Workpiece** | Raw stock dimension, datum | Dimension, shape, roughness |

Figure 1 shows the components that are required for monitoring and control of a machine tool. The sensor module converts mechanical energy into electrical signal that is digitised in a computer. Perhaps the most important element of the monitoring and control system is the signal conditioning and processing module that extracts meaningful information from the raw sensory signals compounded with noises. Subsequently, features correlating to the machining process are obtained through pattern recognition. Finally, a control strategy is made for feedback control.

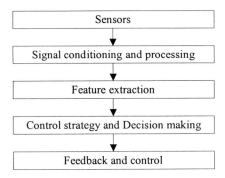

Figure 1 Components of monitoring and control system.

## 2.    Sensors in Machining Process Monitoring

Fundamental to the success of a process monitoring system is the right selection of sensors. A wide variety of sensors have been utilised to monitor machining process. The most important process quantities to be monitored are force, power, acoustic emission and vibration. Much attention has been focused on force measurements and tool condition monitoring (TCM) systems that include tool identification, tool wear monitoring, tool breakage and tool life. Most practical approaches to tool condition monitoring have been developed utilising indirect measurements of tool performance that are easier to achieve than direct measurements. Figure 2 shows the approaches most commonly adopted.

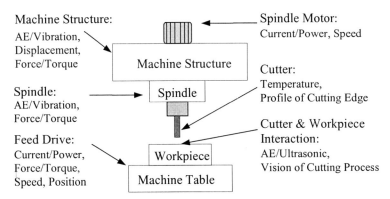

Figure 2 Sensors and machining process monitoring.

## 2.1    *Motor Current and Power*

Spindle motor current or power monitoring is attractive due to its simplicity and non-intrusive nature. A measurement of spindle motor power is related to the load of spindle. Effective power measurement has the advantage over simple current measurement in that there is a linear relationship between change of motor load and that of motor power [2]. In fact the change in motor load is proportional to that in power. However, a large change in motor load may result in a small change in motor current, and the relationship is non-linear. Most often power monitoring is used to prevent overload of the spindle and to detect collisions.

The output signals of these sensors have a low pass filter characteristic for integration calculation of voltage and current. As such tool breakage cannot be detected before it happens, but only after the consequential damage occurs [3]. Besides, it is very difficult to detect tool wear by using current or power sensors. Despite these drawbacks, a tool monitoring system can process these signals to detect a missing tool, a misplaced part, loss of load or overload. The advantage of these sensors is that they are cheap and easy to install on both new and existing machines.

Li, et al [4] used a current sensor installed on the AC servomotor of a CNC turning centre to measure the feed cutting force. A neurofuzzy network was used to estimate feed cutting force based on feed motor current. The experimental results showed that feed cutting force could be accurately estimated using the feed motor current.

Lee, et al [5] investigated an external plunge grinding process with current signals of a spindle motor through a Hall-Effect sensor. By analysing the current signals of the spindle motor, a relationship between current signals and the metal removal rate in terms of the in-feed rate was induced. Research results showed that the current signals of spindle motor reflected the qualitative characteristics of the grinding force. It is possible to predict the metal removal rate by using the current signals of the spindle motor. The authors also compared motor current sensor with an AE sensor and recommended the use of motor current signals rather than AE energy in practical applications.

Huh, et al [6] built a dynamic AC spindle drive model in turning which represented the dynamic relationship between cutting force, motor torque and motor power. The motor power measured during the machining process includes not only the metal cutting torque, but also the non-linear friction torque. The non-linear friction characteristics are identified through off-line cutting tests, while the time-varying effects are compensated with the simple tuning of the damping coefficient. For the steady-state performance, the accuracy of the estimated cutting force is within an error of about 2% actual cutting force in most cutting tests. However, in the transient period the estimated cutting force shows a time lag behind the actual cutting force signal. It is believed that the time lag effect is caused by the modelling error, whereby the synchronous speed of the AC motor is approximated to the motor speed.

In [7], Huh proposed a new synthesised cutting force monitoring method to improve the transient estimation performance. Six signals including motor speed, three voltages and two currents were measured to estimate the cutting force. Based on the cutting force measured in turning, three control strategies of PI (proportional-integral), adaptive and fuzzy logic controllers were applied to investigate the feasibility of utilising the estimated cutting force for turning force control. The experimental results demonstrated that the proposed systems could be easily realised in a CNC lathe with little additional hardware.

## 2.2    *Force/Torque*

Forces are among the most fundamental signals of the machining processes, and one of the most reliable information sources. When a tool applies forces to a workpiece, it results in elastic and plastic deformations in the shear zone and leads to shearing and cutting of the material. The process behaviour is reflected by the changes in the cutting forces, hence monitoring of these quantities are highly desirable. The accurate knowledge of in-process machining forces would significantly benefit process monitoring and control. In general, force measurement is used to monitor tool wear or breakage so as to reduce part damage, or to regulate the machining force for higher material removal rate and longer tool life.

Force measurements are commonly taken by table mounted dynamometers [8]. The cutting forces in three orthogonal directions, the *X*, *Y* and *Z*-axis can be measured. Dynamometers have proved to be a very successful tool for laboratory experimental work. Generally, the cutting forces will increase with end mill tool wear. However, their characteristics vary with changes in cutting conditions and machining direction, in addition to tool wear.

In direct force measurement, the workpiece is mounted to a dynamometer that is in turn fixed to a machine table, as illustrated in Figure

3(a). In such a configuration, the dynamometer is sandwiched between the workpiece and machine table. As a result, the machining envelope in the Z direction is compromised by the size of the dynamometer. This constraint can be overcome by indirect force measurement. In the indirect measurement configuration, the dynamometer is mounted in parallel with the workpiece, as shown in Figure 3(b).

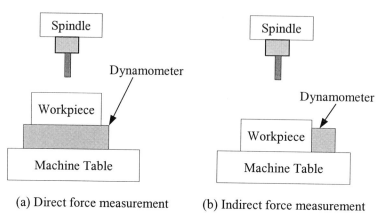

(a) Direct force measurement          (b) Indirect force measurement

Figure 3 Force measurement using dynamometer.

Soliman et al [9] described a control system for chatter avoidance in milling by monitoring the cutting process with a cutting force sensor. A statistical indicator named $R$-value is calculated as the ratio of the root mean squares (RMS) of the low and high frequency components of the cutting force signal. The low and high frequency components of the cutting force signal were obtained by passing the signal through low pass (cut-off frequency 100 Hz) and band pass (250 Hz – 550 Hz) filters. When chatter is detected, the control system ramps the spindle speed in search of a speed at which chatter ceases. The system does not involve time-consuming computations and therefore is suitable for on-line implementation.

A new method, based on a disturbance observer for the reconstruction of the machining forces during cutting, was described by Pritschow [10]. After building a disturbance observer for an electro-mechanical servo drive system, estimation of the process forces is calculated based on the internal variables of the digital servo drive system. The advantage of this method is that it does not need additional sensors. The proposed observer was applied to a laboratory servo drive system and an industrial milling process in a machining centre. The results showed that it is possible to reconstruct the forces that occur between the workpiece and the tool during cutting.

The relationship between the cutting force characteristics and tool usage (tool life or tool wear) in a micro end milling operation was studied for two different metals by Tansel et al [11, 12] with a Kistler dynamometer. Three different encoding methods were used to estimate the tool usage from the cutting forces by using backpropagation (BP) type neural networks. One was a force-variation-based encoding technique that focused on the variation (max-min difference) of the cutting force. The other was a segmental-average-based encoding technique that calculated the segmental average of one rotation. The last one was wavelet-transformation-based encoding (WTBE). In this method, the approximation coefficients of the wavelet transformation were calculated and normalised. They were then fed into a BP neural network and a probabilistic neural network (PNN) respectively. The outputs of these networks were the wear estimation. The best results were obtained when WTBE was used. Training of the BP was almost ten times longer than that of the PNN, but the average estimation error with BP was two to five times better than PNN. Experimental results showed an excellent average wear estimation with the combination of WTBE and BP network, resulting in an error of 7.07% over the entire range of the test data.

## 2.3    *Vibration/Acceleration Signals*

Oscillations of cutting forces lead to vibration of the machine structure. The vibration is also affected by tool wear, as in the case of cutting force. Direct measurement of vibration is difficult to achieve because the vibration mode is frequency dependent. Hence, related parameters such as the rate at which dynamic forces change per unit time (acceleration) are measured and the characteristics of the vibration are derived from the patterns obtained. The acceleration signals are able to provide an earlier indication of approaching cutting failure than the force signals can possibly achieve. The emphasis of research work in this area is signal processing.

Accelerometers are widely used to monitor the cutting process due to their low cost, ease of use, and adaptability. They can be readily mounted onto a machine table or workpiece, typically away from the rotating cutter edge. Accurate measurements of tool wear must take into account the effects of the signal path from the source of excitation at the cutter edge to the measurement site on the machine tool. It is known that the mechanical interfaces in the signal path are especially important. However the transmission of energy across the interfaces is not well understood.

Roth et al [13] presented a system to monitor end mill wear and predicted tool failure using accelerometers. End milling signals are intermittent in nature due to the flutes of the cutter engaging and disengaging with the workpiece. The analysis of the milling process is further complicated by the high level of noise picked up by the accelerometer and the low level of energy associated directly with the cutting parameters. It is necessary to decompose the original signal into its individual contributing modes although both autoregressive (AR) models and autoregressive moving-average models are found to provide reliable results in some cases. This can be accomplished by the Data Dependent Systems methodology using adequate AR models. By monitoring the modal energy components in a small band around the frequency of interest, the modal energies are shown to be closely linked to the wear curve. A detection scheme can be developed to track the end mill's wear and provide an early warning of impending failure.

El-Wardany et al [14] conducted a study on monitoring tool wear and failure in drilling using vibration signature analysis techniques. Vibration signature features, sensitive to the tool wear and breakage, were studied in both time domain (based on ratio of absolute mean value and kurtosis) and frequency domain (based on power spectra and cepstrum ratio). Experimental results showed that the kurtosis values increased drastically with drill breakage, while frequency analysis revealed sharp peaks indicating drill breakage.

## 2.4    *Optical and Vision System*

Optical methods are usually used to measure the profile of the cutting edge or the surface roughness after machining. Detecting tool wear from image processing of the cutting tool has been pursued for many years. Typically, the image of the cutter is captured and analysed to provide information on wear pattern or quantity of wear.

Wong et al [15] devised a vision-based tool condition monitoring system using laser scatter pattern of reflected laser ray in the roughing to near-finish range. A laser source was focused on the finished workpiece such that its reflected ray was captured through a digital camera. The recorded image was processed and characterised using the mean and standard deviation of the scatter pattern and their distribution correlated to the surface roughness. The deduced surface roughness was then related to the state of the cutting tool wear. They concluded that it was very difficult to determine tool wear by observing machined surface roughness.

Kim et al [16] proposed a tool wear monitoring strategy by measuring spindle shaft torsional vibration with an optical system in milling. A light beam acting as a probe was directed onto the spindle shaft whose motion modulated the optical probe. The reflected light beam was analysed to deduce shaft vibration changes. Owing to the shaft motion, the reflected light is frequency shifted and the measured shift in frequency can be used to calculate shaft velocity. The signal was then analysed in the frequency domain, yielding a measurement of spectral power. Results showed that this spectral power ratio was correlated with cutter wear over the range of cutting conditions used.

Optical probing overcomes the limitations of dynamic inertia associated with a mechanical probe. Furthermore, it is able to access constrained areas of a machine by means of an optical fibre.

## 3. Acoustic Emission Sensing

Among the various schemes for process monitoring, detection of Acoustic Emission (AE), generated during the machining operation, is commonly used. AE sensing combines some of the most important requirements for sensor systems, such as, relatively low costs, no negative influence on the stiffness of the machine tool, and ease of mounting.

An AE signal is a low amplitude, high frequency stress wave generated by a rapid release of strain energy in a medium. Metal cutting involves plastic deformation, fracture and friction processes; all of which are rich sources of AE. Application of AE sensing to process monitoring is advantageous because the generation of AE is closely related to basic mechanisms of the cutting process. Changes in these processes lead to those in the AE signal. The close relationship between the cutting process and AE energy makes AE sensors an effective means for process monitoring.

Typically, an acoustic emission sensor operates in the range of 0.1 to 1 MHz. The acoustic signal is amplified and filtered to eliminate extraneous background noise. The signal is analysed to extract relevant parameters. These parameters are either in the time domain, such as total counts, count rate, event counts, event rate, amplitude distribution and energy; or in the frequency domain where the frequency spectra of the signals is related to sources of acoustic emission. Correlation techniques, pattern recognition and statistical approaches are then employed to analyse these parameters.

## 3.1    *Acoustic Emission Mechanism*

Acoustic emission is the elastic energy that is spontaneously released by materials undergoing deformation. It can be formally defined as "the class of phenomena where transient elastic waves are generated by the rapid release of energy from localised sources within a material, or the transient elastic waves so generated" [17]. This definition embraces both the process of wave generation and the wave itself. The stress redistributions are normally caused by the generation of structural changes in a material under a general loading condition. Examples of structural changes are crack growth, phase transformations, corrosion, wear, etc. The condition can be mechanical, thermal or chemical.

Acoustic emission can be considered as the naturally generated ultrasound, created by local mechanical instabilities such as dislocations, micro cracks and phase transformations within a material.

Plastic deformation is the primary source of acoustic emission in loaded metallic materials. The plastic deformation of crystalline solids can be explained on a microscopic level in terms of the motion of dislocations within the medium, under the action of an applied stress. Dislocations are line defects in the crystal lattice which allow slip to occur in ordinary engineering materials at stress levels far below the theoretical shear strength of the crystal lattice. Many researchers have postulated that the motion of dislocations during plastic deformations is responsible for AE generation. However, controversy still exists over the exact details of the mechanism by which dislocation motion produces AE.

Apart from plastic deformation, fracture and friction sources also generate AE. Fracture related AE is attributed to the formation and propagation of micro cracks. As the material becomes strain hardened, a large number of dislocations pile up at grain boundaries, causing back stresses to develop. Ultimately, a point is reached where the action of the applied load and the back stresses at the grain boundaries causes micro crack formation. This generates AE due to plastic deformation and the resulting dislocation motion at the crack tips. As higher stresses are applied, the energy barrier required to create new crack surfaces is overcome, causing the micro cracks to propagate. The propagation of micro cracks releases elastic energy due to the creation of new surfaces, which is detected as AE.

Friction related AE is attributed mainly to plastic deformation of the surface layers, surface asperities and asperity fracture as the surfaces slide past one another. At usual levels of normal loads on sliders, the real contact area is smaller than the apparent contact area, which causes plastic deformation and welding of surface asperities. As the surfaces move past one another, the welded junctions are overcome by shearing and fracture of asperities, which generates AE. Micro cracks form over the contacting surfaces, and separation of particles from the parent material takes place as a consequence of fatigue and micro-cutting. These processes are proposed as AE sources, although the exact mechanisms are not yet well understood on a quantitative basis.

Propagation of the waves through the medium can significantly affect the wave characteristics so that the final wave detected at the surface may have characteristics different from those of the original waveform emitted at the source. Some important effects during wave propagation are dispersion, wave mode changes, and wave attenuation.

## 3.2 *Acoustic Emission in Machining*

In many manufacturing processes, machining forces in tools and materials generate high-frequency acoustic emission. Various investigations in recent years have shown that AE signals can be used to analyse and monitor machining operations. The use of acoustic emission techniques for process monitoring has the potential of ensuring high product quality while minimising the total cost of a product. Processes such as cutting, grinding, forming, and joining all generate acoustic emission for reasons unique to each process. In many cases, the emission can be monitored to characterise the process, to detect defects or process abnormalities in situ, and to detect finish quality. Acoustic emission sources in different processes are shown in Table 2.

Principal areas of interest, with respect to AE signal generation in metal cutting, are in the primary generation zone ahead of the tool where the initial shearing occurs during chip formation. The secondary deformation zone is along the chip/tool rake face interface where sliding and bulk deformation occur. The third zone is along the tool flank face/workpiece surface interface. Finally, the fourth area of interest is associated with the fracture of chips during the formation of discontinuous chips [18,19].

In milling (rotating multiple toothed cutters), the cutting process is discontinuous with varying chip load and relative rubbing-generated noise due to the swarf motion between the tool and workpiece. Thus the basic AE signal, due to the material deformation inherent in chip deformation and chip/tool contact, is complicated by noise and periodic interruptions of the cutting process. This is in contrast to the relatively stationary single point turning generated AE.

Table 2 AE sources in different processes.

| Process | AE sources | Process Conditions | Product Defects |
|---------|-----------|--------------------|-----------------|
| **Machining** | Plastic deformation in shear zone; Chip/tool friction; Chip fracture; Chip tangling; Tool/workpiece friction; Tool fracture. | Tool fracture; Tool wear; Lubrication failure; Chip form; Chatter; | Burr; Surface finish; Dimensional variation. |
| **Grinding** | Plastic deformation in wheel and workpiece contact zone; Grit fracture; Contact/spark out. | Wheel wear; Contact/spark out; Lubrication failure; Chatter; Burning. | Surface finish; Dimensional variation. |
| **Welding** | Arc variation; Metallurgical phase changes; Cracking; Deformation due to thermal distortion. | Arc stability; Shielding gas effectiveness; Metal transfer; Arc on/off. | Cracking Distortion. |

To summarise, the major sources of acoustic emission identified in metal cutting include:

- Plastic deformation of workpiece material in the shear zone.
- Plastic deformation and sliding friction between chip and tool rake surface.
- Sliding friction between workpiece and tool flank surface.
- Collision, entangling and breakage of chips.

In milling, because of the interrupted cutting conditions, additional sources include:

• Shock waves generated at tool entry.
• Sudden unloading and chip break off at tool exit.

In both cases the tool velocity and metal removal rate are significant parameters affecting the energy of the AE signal. Because of the geometry of tool-workpiece interaction, there are additional variations in the AE generated, mainly resulting from chip thickness and tool velocity changes as the tool rotates through the workpiece material.

Acoustic emission is usually one of two distinct types: continuous signals and burst signals. Distinct emission events can be difficult to detect from within the continuous signals that arrive at the transducer in a large number. However, they are readily detectable from within the burst signals. In the machining process, for example, continuous acoustic emission signals are generated in the shear zone, at the chip/tool interface and at the tool flank/workpiece surface interface; while burst signals are generated by the chip breakage during or after chip formation or by insert fracture.

The generation of AE from a machining process may extend over frequencies of several MHz, though the signal intensity is usually very low and diminishes with increasing distance from the source. In the low frequency range these signals are mostly associated with machine vibrations and environmental interferences. As such, AE signal analysis for process monitoring is usually conducted in the frequency range of 50 kHz or above. Generally, interference sources in a machine tool are from the electrical and hydraulic feed drives, rolling-contact bearings, spindles, and gears.

Acoustic information has been associated with the state of the cutting tool for many years. Experienced machinists can detect a change in the cutting conditions or the tool state from the audible sound of the process. In a metal cutting operation, much of AE results from the plastic mechanical processes of crack formation and chip removal, and from surface friction. Tool wear alters the contact surfaces between tool and workpiece, and hence the intensity of AE. If tool wear or catastrophic failure (fracture) occurs the AE signal during cutting is either modified or, in the case of fracture, exhibits a dramatic spike from the energy released at fracture.

A common approach to detect tool wear is to analyse the AE RMS energy. A sudden increase in energy can be observed as the wear land increases. Statistical techniques are applied to the AE RMS energy recorded during progressive tool wear to attempt to minimise the sensitivity of the signal to process parameters for the tracking of tool wear. Advance warning of tool breakage is sometimes given by the appearance of micro cracks in the tool, which cause AE signals. This may be utilised for quality and tool monitoring, or process control.

The major advantage of using acoustic emission, to detect the condition of tool wear, is that the frequency range of the AE signal is much higher than that of the machine vibrations and environmental noise. Therefore, a relatively uncontaminated signal can be obtained using a high-pass filter or a well-designed AE sensor with high-pass frequency response. It does not interfere with the cutting operation, thus allowing for continuous monitoring of tool condition. However, AE signals often have to be treated with additional signal processing schemes to extract the most useful information because of the high sensitivity of the signal to a broad array of sources in addition to the cutting process. The sensitivity of the AE signal to the various contact areas and deformation regions in the cutting and chip formation process has led to the analysis of AE signals as a basic tool to monitor the cutting process.

## 3.3    *Acoustic Emission Sensors*

The term acoustic emission sensor is used to describe the type of transducers conventionally used in non-destructive testing techniques. Typically, an AE sensor is mounted onto a test object's surface to detect dynamic motion resulting from acoustic emission events and to convert the detected motion into a voltage-time electrical signal. This electrical signal, strongly influenced by the characteristics of the sensor, is used for all subsequent analysis in the acoustic emission technique. Since the test results obtained from signal processing depend so strongly on the electrical signal, the type of sensor and its characteristics are important to the success and repeatability of acoustic emission testing.

A wide range of basic transduction mechanisms, such as capacitive transducers, displacement sensors or even the laser interferometers, can be used to detect acoustic emission. Nevertheless, acoustic emission detection is commonly performed with sensors that use piezoelectric elements for transduction. Figure 4 shows the schematic of a typical AE sensor employing a piezoeletric element. The piezoelectric element is usually a special ceramic, such as lead zirconate titanate (PZT), and is acoustically coupled to the surface of the test item so that the dynamic surface motion propagates into the piezoelectric element. The dynamic strain in the element produces a voltage-time signal as the sensor output.

Acoustic emission results from surface motion that consists of a motion component normal to the surface and two orthogonal components tangential to the surface. In principle, acoustic emission sensors can be designed to respond selectively to any motion component. However, virtually all commercial acoustic emission sensors are made to be sensitive to the normal component. Since shear and Rayleigh waves typically have normal motion components, acoustic emission sensors can detect these waves.

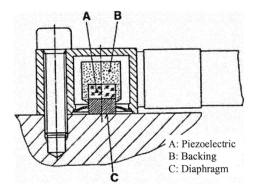

A: Piezoelectric
B: Backing
C: Diaphragm

Figure 4 A typical AE sensor.

Mostly acoustic emission sensing is based on the processing of signals with frequency contents in the range from 50 kHz to about 1 MHz. In some special applications, detection of acoustic emission at frequencies below 20 kHz or near audio frequencies can improve testing, and conventional microphones or accelerometers are sometimes used.

Attenuation of the wave motion increases rapidly with frequency. For a material with higher attenuation, it is necessary to sense lower frequencies in order to detect acoustic emission events. For materials with low attenuation, acoustic emission events are detected at higher frequencies, and the background noise is also lower.

Acoustic emission sensing often requires a couplant between the sensor and test material. The purpose of the couplant is to provide a good acoustic path from the test material to the sensor. Without a couplant or a very large sensor hold-down force, only a few random spots of the material-to-sensor interface will be in good contact, resulting in little energy arriving at the sensor. To sense a normal motion, virtually any fluid like oil, water and silicon grease can act as a good couplant, and the sensor output can often be thirty times higher than that without a couplant. The contact surface must be carefully prepared to ensure complete contact between the sensor and surface, thus avoiding discontinuity in wave transmission. Cleanliness, roughness, and curvature also affect AE transmission. Usually a hold-down force of several Newtons is exerted onto the sensor to ensure good contact and to minimise the couplant thickness. Excessive pressure or lack of pressure, which may introduce errors, must be avoided. For sensing tangential motions, it is more difficult to find a suitable couplant because most liquids do not transmit shear forces. Some high viscosity liquids such as certain epoxy resins are reasonably efficient in sensing tangential motion.

There can be a strong relation between temperature and the piezoelectric characteristics of the active element in an acoustic emission sensor. Some of these effects are important to the use of acoustic emission sensors in testing at elevated or changing temperatures. Typically, a piezoelectric ceramic has a critical temperature, called the Curie temperature, at which the properties of the ceramic change permanently and the ceramics no longer exhibits piezoelectricity. The Curie temperature of PZT ceramics is 300 to 400 °C depending on the type of PZT.

Special problems are encountered when sensors are placed in environments with widely changing temperatures. Temperature changes can cause the direction of electrical polarisation, in some small domains of piezoelectric ceramics, to flip. This results in a spurious electrical signal that is not easy to distinguish from the signal produced by acoustic emission. Typically, a temperature change of 100°C may cause an appreciable number of these domain flips.

Ceramic elements should be allowed to reach thermal equilibrium before using at differing temperatures. If acoustic emission testing must be performed during large temperature changes, then single crystal piezoelectric materials such as quartz are recommended.

A good AE sensor should have a small size so that it is suited for installation near the production process (i.e. the AE source), even where the available space is very confined. This brings optimal avoidance of interference and signal attenuation, assuring wideband transmission of the AE signals. Sensors for process monitoring must meet the following requirements:

- Measurement point as close to the machining point as possible.
- No reduction in the static and dynamic stiffness of the machine tool.
- No restriction of workspace and cutting parameters.
- Wear-free, maintenance-free, easy to change, and low cost.
- Resistant to coolant, dirt, and chips; and to mechanical, electromagnetic and thermal influences.
- Function independent of tool or workpiece.
- Adequate metrological characteristics.
- Reliable signal transmission.

## 4.    Advanced Signal Processing Techniques

Acoustic emission signals are sound waves generated in solid media. They are similar to the sound waves propagated in air and other fluids but are more complex. The AE signals are affected by characteristics of the source, the path taken from the source to the sensor, the sensor's characteristics, and the measuring system. Generally the AE signals are intricate, and using them to characterise the source is difficult. Information is extracted by methods ranging from simple waveform parameter measurement to pattern recognition using artificial intelligence. The former often suffices for simple tests. The latter may be required for on-line monitoring of complex systems. Interpretation of the signals generated during a machining process often requires advanced signal processing.

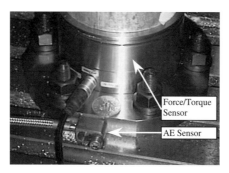

Figure 5 AE sensor mounted on the fixture.

Figure 5 shows the experimental set-up in this study. A force/torque sensor and an AE sensor were deployed. This section only discusses the AE sensor and signal processing. A three-axis end mill CNC machine was deployed to carry out the experimental work. A four-flute end mill cutter was chosen to machine an aluminium alloy material. Figure 6 and Figure 7 show burst AE signals.

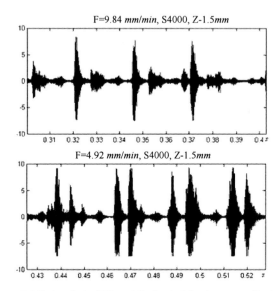

Figure 6 AE signals in different feed rate (abscissa: *s*, ordinate: *v*).

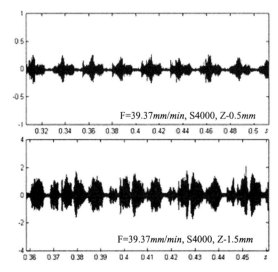

Figure 7 AE signals in different cutting depth (abscissa: *s*, ordinate: *v*).

## 4.1    *Time Domain Analysis*

Discrete or burst type acoustic emission can be described by relatively simple parameters. The signal amplitude is much higher than the background and is of short duration, from a few microseconds to a few milliseconds. Occurrences of individual signals are well separated in time. Although the signals are rarely simple waveforms, they usually rise rapidly to a maximum amplitude and decay nearly exponentially to the level of background noise. In practice, acoustic emission monitoring is carried out in the presence of continuous environmental noise.

Typically, the waveform parameters of an AE signal include an AE event, AE count, AE event energy, AE signal amplitude, AE signal duration and AE rise time, as illustrated in Figure 8.

AE events are individual signal bursts produced by local material changes. The emission count is the number of times a signal crosses a preset threshold. High amplitude events of long duration tend to have many threshold crossings. The number of threshold crossings per unit time depends on the sampling frequency and the threshold level.

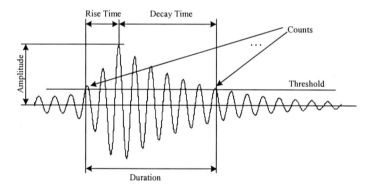

Figure 8 Definition of waveform parameters.

The counts are a complex function of the frequency response of the sensor and the structure, the damping characteristics of the sensor, the event, propagation medium, signal amplitude, coupling efficiency, sensor sensitivity, and amplifier gain. Maintaining stability of these parameters, throughout a test or from test to test, is difficult, but it is essential for consistency of interpretation. Nevertheless, among all parameters, counts are widely used as a practical measure of acoustic emission activity.

Since acoustic emission activity is attributed to the rapid release of energy in a material, the energy content of the acoustic emission signal can be related to this energy release and can be measured in several ways. Direct energy analysis can be performed by digitising and integrating the waveform signal. In the case of continuous emission, if the signal is of approximately constant amplitude and frequency, the energy rate is the root mean square voltage (RMS), which is defined as:

$$RMS = \sqrt{\frac{1}{N}\sum_{i=1}^{N} x_i^2}$$

where $x_i$ is the sampled voltage output.

The RMS voltage measurement is simple, but RMS response is generally slow in comparison with the duration of most acoustic emission signals. Therefore, RMS measurements are indicative of the average acoustic emission energy rather than the instantaneous energy measurement of the direct approach. Regardless of the type of energy measurement used, none is an absolute energy quantity. They are relative quantities proportional to the true energy.

Pruitt et al [20] described a method to monitor end mill contact using acoustic emission sensing. The acoustic signals were processed by a root mean square meter to provide a measure of acoustic emission energy. A detectable increase in AE RMS occurs in a region 20 to 65 microns before the material being removed. This AE signal is produced by workpiece interaction with the turbulence in the boundary layer of the end mill and by elastic contact of the tool and workpiece. Although the accuracy of the tests, discussed in this chapter, was not sufficient for part location for precision machining or measurement purposes, the AE RMS signal can still be monitored in process to increase the pre-material-removal feed rates.

Beggan et al [21] used acoustic emission to predict surface quality in turning and found that a positive correlation existed in the measured values of AE RMS and the actual surface roughness. This correlation can be employed to predict the surface roughness of a workpiece while it is still in production.

The milling process exhibits a periodic generation mechanism, perturbed by spindle speed variations, tool conditions, and the effect of chips. The ideal AE RMS signal during milling varies periodically with a frequency equal to the product of spindle rotational frequency and the number of independent cutting edges. As cutting time increases, tool wear during end milling is developed mainly on the flank faces of the helical flutes. It is usually observed that the wear lands on different flutes are not the same, and this leads to variation in chip thickness as wear increase. Thus the actual AE RMS signal can be regarded as a periodic random signal. With increasing flank wear, the mean AE RMS increases as does the variation of the mean signal from the average periodic position. The latter is attributed to non-uniform wear distributions on the separate cutting edges. The mean AE RMS is highly dependent on the milling parameters and is thus not suitable for application to in-process monitoring of tool wear.

According to this concept, a signal processing scheme, based on time domain averaging (TDA), has been developed by Hutton et al [22]. The normalised mean TDA deviation is proposed as a characteristic of the AE RMS signal to track tool wear during milling. The mean TDA deviation describes the dynamic component of the AE RMS signal and intrinsically relates to the features of tool wear in milling. Experimental data show that the normalised mean TDA deviation is sensitive to flank wear, but less dependent on cutting parameters than the AE RMS signal is.

Due to the stochastic nature of the AE signals generated in the metal cutting process, the analysis of the signals has been often approached statistically. One of the methods is based on the distribution moments of the AE signal. The kurtosis of a signal is defined as the ratio of the fourth moment of distribution to the square of the second moment. It is a measurement of the sharpness of the peak. A high kurtosis value implies a sharp distribution peak (concentration in a small area) while a low kurtosis value indicates essentially flat characteristics. Kurtosis is useful in identifying transients, or spikiness, in a signal. Webster et al [23] found that the transients caused by the initial contact between grinding wheel and workpiece could be clearly identified much earlier than with AE RMS filtering. The kurtosis was calculated over a sliding exponential window.

In general, measurements of events, counts, and energy provide an indication of source intensity. These are information useful for determining whether the test object is accumulating damage. In many cases, the high sensitivity of acoustic emission sensing also detects unwanted background noises that cannot be removed by signal conditioning. In spite of many attempts to establish fundamental relationships between the simple parameters, correlations between signal and source by analytical methods remain elusive. This dilemma is a result of the complexity of modelling the source, the material, the structure, the sensor and the measurement system.

Advanced signal processing methods provide an alternative, permitting successful characterisation and source identification.

## 4.2    *Time Series Modelling*

An important factor to consider when modelling the machining process is that the process can be divided into two independent models: a deterministic model and a stochastic model. It is sometimes possible to derive the mathematical model for a dynamic system based on physical laws. This type of model is entirely deterministic, and allows the calculation of the value of some time-varying variables at any particular time. For machining modelling, it is based on known dynamic characteristics of the machining process such as radial depth of cut, axial depth of cut, spindle speed, number of cutting teeth and spindle position.

However, very few dynamic systems, especially those in manufacturing, are totally deterministic because changes due to unknown or un-quantified effects may take place during the process. Thus, it is often convenient to construct stochastic models that can describe the dynamics from a probabilistic point of view. In this way the underlying system physics or system characteristics can be studied or determined from experimental data. Tool wear and breakage would also be considered as stochastic factors.

One method frequently utilised is time series modelling. A time series is a sequence of observed data ordered in time. Many algorithms in the time domain have been developed to estimate the model of time series data. Two main time series modelling techniques have been used commonly: autoregressive time series and autoregressive moving average time series (ARMA). Time series analysis, which utilises the correlation of successive observations of each informative feature, is often coupled with a pattern recognition method.

Liang [24] utilised low order autoregressive (AR) models with a stochastic gradient algorithm for modelling the behaviour of AE RMS energy signals during turning. The aim was to determine progressive flank wear in the presence of large variations in process parameters such as cutting speed, feed rate and depth of cut. This method does not require any prior knowledge of the signal statistical properties and the AR parameters are adaptively updated using a stochastic gradient algorithm.

A new technique, the tooth period modelling technique (TPMT), was proposed by Tansel et al [25] for tool breakage monitoring during milling operations. TPMT combines the advantages of time series based methods and tooth period average cutting force evaluation techniques. It samples the data at fixed spindle rotations to separate the tooth periods, and models the signal with time series AR model techniques to accurately estimate the characteristics of the signal. Tool breakage is detected by monitoring the sum of the squares of the estimation errors of tooth periods that fluctuate periodically when a tooth of the cutter is completely broken or starts to remove less material than the other teeth.

## 4.3    *Frequency Domain Analysis*

Since the AE RMS signal is a statistical average, it is equivalent to a low-pass filtered signal. Some important instant features, for example, high frequency chatter, minor dynamic geometric error of the workpiece and initial contact of cutter and workpiece, may not be revealed through these signals. On the other hand, there is no general guideline for selecting a suitable bandwidth and integration time constant for AE RMS measurement. Therefore, a fundamental study of the frequency characteristics of the AE signal is very important for effectively applying the AE technique for reliable and cost-effective condition monitoring.

Frequency characteristics are analysed with the Fast Fourier Transform (FFT) when they are needed. FFT and spectral analysis are the most commonly used signal processing techniques in engineering. Especially after the development of special FFT hardware based on vector processors, FFT has been applied to many applications. Spectral analysis seeks to describe the frequency content of a signal, random process, or system, based on a finite set of data. Estimation of power spectra is useful in a variety of applications, including the detection of signals buried in wide-band noise.

Hundt [26,27] employed spectral analysis to study AE signals in grinding. After data acquisition and digitisation, time domain segments were transformed to the frequency domain using a hamming window to avoid leakage. The transformed characteristics were then smoothed by a moving average algorithm. These spectra were least square approximated by multiple order polynomials, the polynomial coefficients serving as features. The features of every single event were then mapped in a multi-dimensional feature space, where similar event types will form clusters. Experiments showed that this kind of technique can separate and describe several different AE sources corresponding to different material removal rates. However, their correlation to physical events was still under further research.

Vibration signature features sensitive to the tool wear and breakage were studied in both time domain and frequency domain [14]. The ratio of the calculated cepstra in the transverse and thrust directions was used as a monitoring index for detecting drill breakage. The instantaneous Ratio of the Absolute Mean Value ($RAMV_i$) was used to start cepstrum calculations. The area under the power spectrum curve was used to monitor the wear of large size drills (6 mm in diameter). The $RAMV_i$ was used to trigger the calculation of the vibration power spectrum generated during the drilling of a specified number of holes, where the change in the wear state was minimum. A good correlation was found between the power spectra calculated and the tool wear.

The frequency analysis of the raw AE signal was carried out based on the knowledge of the frequency response of the sensor employed. Research showed that the frequency character of AE signal was usually dominated by the frequency response of the sensor employed. In some cases the frequency character of the AE source can be detected by observing the shift of the frequency position of the main AE energy components from the major resonant frequency to a minor resonant frequency of the sensor. The sensor employed here was a resonant type, with major resonant frequency of 330 kHz [28].

## 4.4    *Time-Frequency Domain Analysis*

In frequency domain analysis, the basic approach to process monitoring with AE signals, is to correlate the FFT analysis spectrum with machining process parameters or tool wear. Tool wear can be detected and identified with this method through an extensive programme of experimentation and data analysis.

However, there are some limitations in this approach. A condition for FFT analysis is that the signal should be stationary and time-invariant. This implies that the process must also be stable. If there exists a transition during the process, FFT would not be appropriate. It is, however, this very transition during the machining process, that is of great interest to a process monitoring engineer, i.e. a non-stationary, time-variant phenomenon.

While the FFT can show what frequencies are contained in the signal, it does not show how these frequencies are varying as a function of time. It is very important for process control to preserve time domain information of signals after they are analysed. In FFT analysis, the basic function $f = e^{i\omega x}$ is poorly localised in time, although its Fourier transform $\hat{f}$ is perfectly localised in frequency at point $\omega$. It is difficult for a FFT to analyse complex signals such as transients or abrupt changes that are the symptoms of defects.

A more sophisticated approach to machining process monitoring is to analyse the time-frequency spectrum of the signal for patterns corresponding to the process characteristics of interest. The ability to detect these time-frequency characteristics is important, since the distribution of frequencies provides information about the process status. This approach has a potential advantage of insensitivity to signal intensity variations due to noise or other similar problems. By analysing the time-frequency spectrum of signals, a larger amount of information can be extracted than by investigating frequency spectrum only. Wavelet analysis is one such time-frequency analysis method.

Suh et al [29] developed a diagnosis method which used the scalogram and the modulus maxima representation of wavelet analysis. The vibration signals were evaluated by the continuous wavelet transform to analyse and visualise the underlying characteristics of the signals in the time-frequency domain. A fault severity index (*FS*), which quantified the leakage of energy into a specific frequency region $R1$ (500 ~ 800 Hz) with time, was defined as the ratio of the signal energy in $R1$ to the total signal energy. If the value of *FS* crosses the thresholds, it indicates that a defect in a gear is severe and that appropriate maintenance actions need to be taken immediately. The modulus maxima of each detail output of the wavelet transform were fed into a feedforward neural network to classify the current state of the gearbox into 4 different classes with a high success rate of 93%.

The acoustic emission signal from the cutting process consists of continuous and transient (or burst) signals, each with distinct characteristics. The transient AE signal is non-stationary because its short impulse type burst is both frequency and magnitude modulated. Niu et al [30] processed the transient AE signal with best-basis wavelet packet transform. For transient tool condition categorisation, a sixteen-element feature vector, corresponding to the frequency band value of wavelet packet coefficients in the phase plane at the burst location, was extracted. This was used to distinguish tool fracture, chipping and chip breakage by using an ART2 (adaptive resonance theory) neural network. Experimental results had shown that this multi-category identification approach was highly successful over a range of cutting conditions.

Suzuki [31] used wavelet transform to analyse acoustic emission signals. AE signal features were easily classified into four patterns through 2D contour maps and 3D projections of wavelet coefficients. The main drawback of the wavelet analysis is the length of computation time in comparison with the nearly instantaneous calculation of FFT.

Ziola [32] discussed several digital signal processing methods of AE signals. Several transforms were tested with varying degrees of success. The Short Time Fourier Transform (STFT) is computationally fast, but lacks the frequency resolution required for the analysis. Wavelet and Wigner-Ville transforms provide better frequency resolution, but require a long time to perform the computations. To shorten the computation time, Gaussian cosine signals of various frequencies were cross-correlated with the signal, and the envelopes of the resulting cross correlations were plotted as a function of frequency. While this method is not mathematically rigorous, it does provide the necessary frequency resolution, while still being computationally fast.

Bukkapatnam et al [33] combined chaos theory, for thorough AE signal characterisation, with wavelet theory, for signal representation and modelling. Their approach, based on their earlier methodologies of characterising and modelling turning dynamics, resulted in a compact time-frequency domain representation of AE signals. From this representation, they developed an adequate neural network based tool wear estimation scheme.

## 4.5  *Wavelet Analysis*

According to the Heisenberg Uncertainty Principle, the product of signal dispersion $f$ and its Fourier Transform dispersion $\hat{f}$ is always greater than a constant $c$, which does not depend on the signal but only on the dimension of the space. Therefore, it is impossible to reduce both time and frequency localisation arbitrarily. Wavelet analysis provides an interesting compromise on this problem. Applying windows with different sizes can change the resolution of time and frequency. Wavelet analysis allows the use of long intervals when more precise and low frequency information is needed, or the use of shorter intervals when high frequency information is needed.

A wavelet is a waveform with effectively limited duration and zero average value. If the given wavelet function is $\psi(t) \in L^2(R)$, wavelets are defined as:

$$\psi_{ab}(t) = |a|^{-\frac{1}{2}} \psi(\frac{t-b}{a}) \tag{1}$$

where $a$ is a scaling factor, $b$ is a shifting factor, and $L^2(R)$ is the set of signals of finite energy.

The continuous wavelet transform (CWT) is defined as the sum over all time of the signal multiplied by scaled, shifted versions of the wavelet function:

$$(W_\psi f)(a,b) = |a|^{-\frac{1}{2}} \int_R f(t) \, \overline{\psi}(\frac{t-b}{a}) \, dt \tag{2}$$

where $a,b \in R$, $a \neq 0$ , $R$ is the set of real numbers, and $\overline{\psi}(t)$ is conjugate function of $\psi(t)$.

Calculating wavelet coefficients at every possible scale is arduous and time consuming. If scales and positions are based on powers of two, the analysis will be much more efficient. A fast wavelet decomposition and reconstruction algorithm was developed by Mallat [34] in 1988. Mallat's algorithm for discrete wavelet transform (DWT) is a classical scheme in the signal processing community. It is well known as a two-channel sub-band coder, using conjugate quadrature filters or quadrature mirror filters (QMF). This very practical filtering algorithm yields a fast wavelet transform.

One of the fundamental relations of an orthogonal wavelet is the twin-scale relation (dilation equation or refinement equation) described by the following equation:

$$\phi(t) = \sqrt{2} \sum_{n \in Z} h_n \phi(2t - n) \tag{3}$$

where $\phi(t)$ is called scaling function and associated with wavelet function $\psi(t)$, and $Z$ is the set of integer numbers.

All the filters used in DWT and inverse DWT (IDWT) are related to the sequence $(h_n)_{n \in Z}$. If $\phi(t)$ is compactly supported, the sequence $(h_n)$ is finite and can be viewed as a filter $h$. The filter $h$, known as the scaling filter, is a low-pass filter with a length of $2N$, sum of $\sqrt{2}$ and norm of 1. Written with the inner product:

$$
\begin{aligned}
h_k &= <\phi_{j+1,0}, \ \phi_{j,k}> \\
g_k &= <\psi_{j+1,0}, \ \phi_{j,k}>
\end{aligned}
\qquad (4)
$$

where $g$ and $h$ are quadrature mirror filters.

Given a discrete sampling series $f(n)$ $(n = 1, \cdots, N)$ of signal $f(t)$ and denoting the approximation vector of signal at scale $j = 0$ as $C_0(n) = C_j(n)\big|_{j=0} = f(n)$, the dyadic discrete wavelet transform can be expressed as:

$$
\left.
\begin{aligned}
C_{j+1}(n) &= \sum_{k \in Z} \overline{h}(k - 2n)C_j(k) \\
D_{j+1}(n) &= \sum_{k \in Z} \overline{g}(k - 2n)C_j(k)
\end{aligned}
\right|
\qquad (5)
$$

The wavelet decomposition of signal $f(t)$ can be written as:

$$
\begin{aligned}
A_j f(t) &= A_{j+1}f(t) + D_{j+1}f(t) \\
&= \sum_n C_{j+1}(n)\phi_{j+1,n}(t) + \sum_n D_{j+1}(n)\psi_{j+1,n}(t)
\end{aligned}
\qquad (6)
$$

$A_j f$ is the output of applying a low-pass filter to $f(t)$ and is called the approximation. $D_j f$ is the output of applying a series of width-variable band-pass filters to $f(t)$ and is called the detail.

The key point of wavelet analysis is to extract information from the original signal by decomposing it into a series of approximations and details distributed over different frequency bands. The characteristics of frequency domain and time domain are preserved simultaneously. Further processing is then carried out after selecting several decomposition sequences suitable for the given application. The decomposition can be described as a decomposition tree shown in Figure 9.

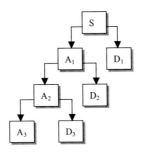

Figure 9 Wavelet decomposition tree.

From the frequency structure of wavelet decomposition, the frequency bandwidth of approximation and detail of level $l$ are $\left[0, \frac{1}{2}f_s 2^{-l}\right]$ and $\left[\frac{1}{2}f_s 2^{-l}, \frac{1}{2}f_s 2^{-(l-1)}\right]$ respectively. It is noticed that the frequency band of every level is decomposed into two equal sub-bands, the detail and the approximation. The result of wavelet translation is a series of decomposed signals belonging to different frequency bands.

The above methodology was applied to analyse the AE signals obtained in a milling operation. Taking the signals in Figure 7 as an example, signals are decomposed to level 6 with a *db5* wavelet. The sampling frequency ($f_s$) is 800 kHz. According to the wavelet decomposition method, the acquired AE signal S (400 kHz) is decomposed into those in the five frequency bands as follows:

D1: [200 kHz,   400 kHz]
D2: [100 kHz,   200 kHz]
D3: [50 kHz,   100 kHz]
D4: [25 kHz,   50 kHz]
D5: [12.5 kHz,   25 kHz]
A5: [0,   12.5 kHz]

The signal energy in *D1* is found to be very low, and hence neglected. Signal energy in the other four frequency bands (D2 to D5) is plotted in Figure 10. It is shown from the figure that the signal energy is mostly distributed over *D2* and *D3* frequency bands that have a frequency range from 50 kHz to 200 kHz. The signals in *D2* and *D3* are much stronger than those in other frequency bands. Hence, information is extracted from these two frequency bands with FFT or other signal processing techniques. For the same reason, information related to tool condition will be concealed and difficult to distinguish if it does not belong to these frequency bands. As shown in Figure 10, the energy of D5: [12.5 kHz, 25 kHz] is very weak. With the help of wavelet analysis, however, the peaks in D5 were revealed; indicating that one flute of the cutter may fracture. It would be very hard to detect such peaks in a weak signal without resorting to wavelet analysis.

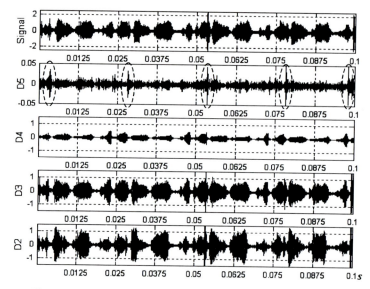

Figure 10 Wavelet analysis of AE signals (abscissa: *s*, ordinate: *v*).

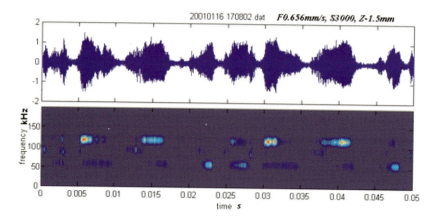

Figure 11 Short time Fourier transform analysis of AE signals.

The AE signals are also analysed with Short Time Fourier Transform (STFT) as shown in Figure 11. The brightness in the figure indicates the intensity of the power spectral density (PSD). The higher the brightness is, the stronger the PSD in the frequency band. The interaction between the four flutes of the cutter and the material can be seen clearly around the frequency 150 kHz. However we could not find any useful information below 50 kHz.

## 5. Conclusions

Process monitoring has evolved extensively over the years, with significant advances having been made in several areas. Despite these achievements, there are still areas that require continued work to further enhance the capability of the process monitoring. Among these are process modelling, open architecture/multi-sensor fusion, signal processing/image processing, and feature extraction.

A fundamental understanding of the process behaviour can be acquired with the help of process modelling. It also helps in the selection of appropriate sensors for a given operation, and to obtain better interpretation of the sensor outputs.

A wide variety of sensors such as acoustic emission sensor, motor current/power sensor, force/torque sensor, vibration/acceleration sensor and optical/vision sensor have been utilised to monitor machining process. Acoustic emission sensor signals exhibit good correlation to cutting condition. The outputs of these different sensors may be complementary or conflicting. In this situation, sensor fusion is needed to coordinate the outputs of different sources of sensory signals.

Signals are easy to be contaminated during transmission in machining environments. As a result, sensory signals are often accompanied with a lot of additional confusing data. In order to provide an accurate interpretation or feature extraction of the information obtained, advanced signal processing and analysis is needed. The feature is, ideally, free of non-useful information and represents a core of information that is sensitive to the process and its behaviour. Today researches are focusing on four signal processing techniques: time domain analysis, time series modelling, frequency domain analysis and time-frequency domain analysis.

Although various sensors have been developed for process monitoring, a reliable monitoring scheme also relies on accurate interpretation of the sensor outputs, and consequently, the relationship between the outputs and the process. It is the basis on which successful and reliable monitoring of the processes can be accomplished. This requirement calls for the development of appropriate signal processing algorithms to analyse the data generated and sophisticated feature extraction algorithms to recognise process conditions. Apart from the signal processing techniques covered in this chapter, pattern recognition and neural networks have good potential in this area.

# References

1.  Chen, X.Q., Zeng, H., and Wildermuth, D. "In-process monitoring through acoustic emission sensing", Gintic Technical Report AT/01/014/AMP, 2001.

2.  Jemielniak, K., "Commercial tool condition monitoring systems", International Journal of Advanced Manufacturing Technology, Vol. 15, 1999, pp. 711-721.

3.  Byrne, G., Dornfeld, D., Inasaki, I., et al., "Tool condition monitoring (TCM) – The status of research and industrial application", Annals of the CIRP, Vol. 44, No. 2, 1995, pp. 541-567.

4.  Li, X., Venuvinod, P.K. and Chen, M.K. "Feed cutting force estimation from the current measurement with hybrid learning", International Journal of Advanced Manufacturing Technology, Vol. 16, 2000, pp. 859–862.

5.  Lee, E.S., and Kim, N.H. "A study on the machining characteristics in the external plunge grinding using the current signal of the spindle motor", International Journal of Machine Tools & Manufacture, Vol. 41, 2001, pp. 937–951.

6.  Huh, K., Jung, J.J., and Lee, K.K. "A cutting force monitoring system based on AC spindle drive", Proceedings of the American Control Conference, Philadelphia, Pennsylvania, June 1998, pp. 3013-3017.

7.  Huh, K., and Kim, J. "Turning force control systems based on the estimated cutting force signals", Proceedings of the American Control Conference, San Diego, California, June 1999, pp. 684-688.

8.  Prickett, P.W., and Johns, C. "An overview of approaches to end milling tool monitoring", International Journal of Machine Tools & Manufacture, Vol. 39, 1999, pp. 105-122.

9.  Soliman, E., and Ismail, F., "A control system for chatter avoidance by ramping the spindle speed", Transaction of ASME, Journal of Manufacturing Science and Engineering, Vol. 120, No. 11, 1998, pp. 674-683.

10. Pritschow, G., Bretschneider, J., and Fritz, S. "Reconstruction of process forces within digital servodrive systems", Institute for Control Technology (ISW), University of Stuttgart, 1999.

11. Tansel, I.N., Arkan, T.T., Bao W.Y., et al. "Tool wear estimation in micromachining. Part I: tool usage cutting force relationship", International Journal of Machine Tools & Manufacture, Vol. 40, 2000, pp. 599-608.

12. Tansel, I.N., Arkan, T.T., Bao W.Y., et al. "Tool wear estimation in micromachining. Part II: neural-network-based periodic inspector for non-metals", International Journal of Machine Tools & Manufacture, Vol. 40, 2000, pp. 609-620.

13. Roth, J.T. and Pandit, S.M. "Monitoring end-mill wear and predicting tool failure using accelerometers", Transaction of ASME, Journal of Manufacturing Science and Engineering, Vol.121, No.11, 1999, pp. 559-567.

14. El-Wardany, T.I., Gao, D., and Elbestawi, M.A. "Tool condition monitoring in drilling using vibration signature analysis", International Journal of Machine Tools & Manufacture, Vol. 36, No. 6, 1996, pp. 687-711.

15. Wong, Y.S., Nee, A.Y.C., Li, X.Q., et al. "Tool condition monitoring using laser scatter pattern", Journal of Materials Processing Technology. Vol. 63, 1997, pp. 205-210.

16. Kim, S., and Klamecki, B. "Milling cutter wear monitoring using spindle shaft vibration", Monitoring and Control for Manufacturing Process, at the Winter Annual Meeting of the American Society of Mechanical Engineers, Dallas, Texas, Nov. 25-30, 1990, pp. 57-74.

17. Miller, R.K., and McIntire, P. "Non-Destructive Testing Handbook, Vol. 5 Acoustic Emission Testing", 2nd Edition, USA: American Society for Nondestructive Testing, 1987.

18. Dornfeld, D.A., "Application of acoustic emission techniques in manufacturing", NDT & E International, Vol. 25, No. 6, 1992, pp. 259-269.

19. Dornfeld, D.A., "In process recognition of cutting states", JSME International Journal, Series C: Dynamics Control Vol. 37, No. 4, 1994, pp. 638-650.

20. Pruitt, B.L., and Dornfeld, D.A., "Monitoring end mill contact using acoustic emission", Proceedings of the Japan/USA Symposium on Flexible Automation, Vol. 1, 1996, pp. 421-426.

21. Beggan, C., Woulfe, M., Young, P., and Byrne, G. "Using acoustic emission to predict surface quality", International Journal of Advanced Manufacturing Technology. Vol. 15, 1999, pp. 757-742.

22. Hutton, D.V., and Hu, F. "Acoustic emission monitoring of tool wear in end-milling using time-domain averaging", Transaction of ASME, Journal of Manufacturing Science and Engineering, Vol. 121, No. 2, 1999, pp. 8-12.

23. Webster, J., Marinescu, I., and Bennett, R. "Acoustic emission for process control and monitoring of surface integrity during grinding", Annuals of the CIRP, Vol. 43, No. 1, 1994, pp. 299-304.

24. Liang, S. and Dornfeld, D.A. "Tool wear analysis using time series analysis of acoustic emission", Transaction of ASME, Journal of Engineering for Industry, Vol. 111, No. 3, 1989, pp. 199-205.

25. Tansel, I.N., and McLaughlin, C. "Detection of tool breakage in milling operations – I. the time series analysis approach", International Journal of Machine Tools & Manufacture, Vol. 33, No. 4, 1993, pp. 531-544.

26. Hundt, W., Leuenberger, D., Rehsteiner, F., et al. "An approach to monitoring of the grinding process using acoustic emission (AE) technique", Annuals of the CIRP, Vol. 43, No. 1, 1994, pp. 295-298.

27. Hundt, W., Kuster, F., and Rehsteiner, F. "Model-based AE monitoring of the grinding process", Annuals of the CIRP, Vol. 46, No. 1, 1997, pp. 243-247.

28. Webster, J., Dong, W.P., and Lindsay, R. "Raw acoustic emission signal analysis of grinding process", Annals of the CIRP, Vol. 45, No. 1, 1996, pp. 335-340.

29. Suh, J.H., Kumara, S.R.T., and Mysore, S.P. "Machinery fault diagnosis and prognosis: Application of advanced signal processing techniques", Annals of the CIRP, Vol. 48, No. 1, 1999, pp. 317-320.

30. Niu, Y.M., Wong, Y.S., Hong, G.S., et al. "Multi-category classification of tool conditions using wavelet packets and ART2 network", Transaction of ASME, Journal of Manufacturing Science and Engineering, Vol. 120, No. 11, 1998, pp. 807-816.

31. Suzuki, H., Kinjo, T., Hayashi, Y, et al. "Wavelet transform of acoustic emission signals", Journal of Acoustic Emission, Vol. 14, No. 2, 1996, pp. 69-84.

32. Ziola, and Steve, "Digital signal processing of modal acoustic emission signals", Journal of Acoustic Emission, Vol. 16, No. 1-4, 1998, pp. 12-18.

33. Bukkapatnam, S.T.S., Kumara, S.R.T. and Lakhtakia, A. "Analysis of acoustic emission signals in machining", Transaction of ASME, Journal of Manufacturing Science and Engineering, Vol. 121, No. 11, 1999, pp. 568-576.

34. Mallat, S. "A theory for multi-resolution signal decomposition: The wavelet representation", IEEE Transactions on Pattern Analysis and Machine Intelligence. Vol. 11, No. 7, 1989, pp. 674-693.

# CHAPTER 5

# TECHNIQUES OF AUTOMATIC WELD
# SEAM TRACKING

Rajagopalan Devanathan*, Sai Piu Chan*, and XiaoQi Chen**

*School of Electrical & Electronic Engineering,
Nanyang Technological University, Nanyang Avenue, Singapore 639798

**Gintic Institute of Manufacturing Technology,
71 Nanyang Drive, Singapore 638075

## 1.    Introduction to Weld Seam Tracking

### 1.1    *The Importance of Welding*

Welding is one of the most economical and efficient ways to join metals permanently. It is a common way of joining two or more pieces of metal to make them act as a single piece or monolithic structure. Welding is used to join all of the commercial metals and to join metals of different types and strengths. Welding is vital to the economy. It is often said that up to 50% of the gross national product of a country is related to welding in one way or another [1,2].

Welding is a very important technique used in manufacturing. It is also a very important technique used in construction, as almost everything made of metal is welded. The use of welding is still increasing. Arc welding equipment represents approximately half of total welding equipment, and arc welding is expected to grow at a rate of 6% annually.

### 1.2    *What is Welding?*

Welding is a joining process that produces coalescence of materials by heating them to the welding temperature, with or without the application of pressure or by the application of pressure alone, and with or without the use of filler metal. A weld is a localised coalescence of metals or non-metals produced through welding. Coalescence is the growing together or growth into one body of the materials being welded.

A weldment is an assembly of component parts, joined by welding. It can be made of many or a few metal parts whose compositions may differ. The parts may be in the form of rolled shapes, sheet, plate, pipe, forgings, or castings. To produce a usable structure or weldment, weld joints are made between the various pieces of the weldment. There are five basic types of joints for bringing two members together.

- Butt joint: two parts in approximately the same plane.
- Corner joint: two parts located approximately at right angles to each other.
- T joint: parts at approximately right angles, in the form of a T shape.
- Lap joint: between overlapping parts in parallel planes.
- Edge joint: between the edges of two or more parallel parts.

The technology of welding is complex. Welding is continuing to grow, yet the industry is changing rapidly. Among these, gas metal arc welding (GMAW) and flux-cored arc welding (FCAW) have become the most popular methods of arc welding.

GMAW is an arc welding process that uses an arc between a continuous filler metal electrode and the weld pool. The process is used with shielding from an externally supplied gas and without the application of pressure. It was developed in the late 1940s for welding aluminium and has become very popular. This process is also called metal inert gas (MIG) welding. There are many variations depending on the type of shielding gas, the type of metal transfer, the type of metal welded, and so on. These include, for example, MIG welding, $CO_2$ welding, fine wire welding, spray arc welding, pulse arc welding, dip transfer welding, and short-circuit arc welding.

FCAW is an arc welding process that uses an arc between a continuous filler metal electrode and the weld pool. The process is used with shielding gas from a flux contained within the tubular electrode with or without additional shielding from an externally supplied gas, and without the application of pressure. This is a variation of GMAW and is based on the configuration of the electrode.

## 1.3    *Automated Welding*

The demand for improved weld quality, reduced welding cost, and increased productivity continues, especially in view of the improved materials and fabricating methods. Automated robotic arc welding is no longer a metalworking curiosity but has become commonplace in the metalworking industry. It plays an important role in industries to speed up reliable fabrication work, as it produces high quality and consistent welds that pass the strictest radiographic checks. This has helped take the human welder away from the arc and fumes. It has helped to clean up the welding environment. Yet, the automation of welding has lagged behind the automation of other industrial production processes. This is because welding is more complex than many other industrial processes [2].

## 1.4    *Need for Seam Tracking*

As we know, the arc welding processes involve heat. High temperature heat is largely responsible for welding distortion, warpage, and stresses. When metal is heated it expands in all directions. When metal cools it contracts in all directions too. Distortion usually occurs in six main forms, such as, longitudinal shrinkage, transverse shrinkage, angular distortion, bowing and dishing, buckling, and twisting. Thus, the seam or the pre-programmed path for automated welding may be distorted once welding is in progress. Hence, there is a need to detect the effect of these distortions and track the actual seam path rather than the pre-programmed path. This requires on-line seam detection and correction (Figure 1).

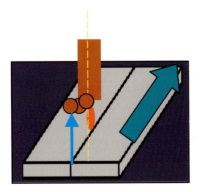

Figure 1 Weld seam.

The term "seam tracking" refers to a low-level control mechanism wherein the welding torch is precisely controlled during welding to position it on the weld seam irrespective of variations in the unstructured seam environment due to thermal distortion, fixture error, improperly prepared weld joints and other such causes.

There are quite a few ways to do seam tracking, such as using mechanical, electrical, sonic, magnetic, or optical sensors [3, 4]. However, two techniques that have received most attention are the Through-the-Arc sensing and optical sensing which will be discussed next.

## 2.     Through-the-Arc Sensing

### 2.1     *Survey of Existing Methods*

The principle of through-the-arc sensing is more than 20 years old [5]. It makes use of the simple fact that, in arc welding using a constant voltage power source and a constant wire feed speed, the resulting current is related to the contact-tip-to-workpiece distance (CTWD). This method [6] utilises the functional relationship that exists between the CTWD and the electrical arc signals. When CTWD increases, there results an instantaneous increase of the arc length and arc voltage and a decrease in arc current. The wire feed rate then no longer equals the melting rate and the system acts to re-establish the equilibrium. Assuming a constant potential self-regulating system, the operating point will return to within a few percent of the voltage and the current values that existed prior to the abrupt change in the CTWD.

Consider, for example, welding on a vee-groove joint as shown in Figure 2. Suppose the torch is oscillated about the programmed path as shown in Figure 3. There are two issues, which need to be considered. One is the height control of the torch or CTWD control. The other is the control of lateral deviation of torch position. An increase in CTWD results in a decrease of current. Compared to a given offset value in a lateral deviation, a simple sensor for height control may be realised by noting that

$$I_{left} + I_{right} = 2 \times I_{preset} \tag{1}$$

where $I_{left}$ and $I_{right}$ correspond to the welding current in the leftmost and the rightmost positions of the torch (see Figure 3) respectively. $I_{preset}$ represents a preset current value. If the left-hand side of Equation (1) is different from twice the preset value, then corrective action to the CTWD is made until Equation (1) is satisfied.

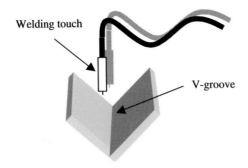

Figure 2 Vee-groove joint.

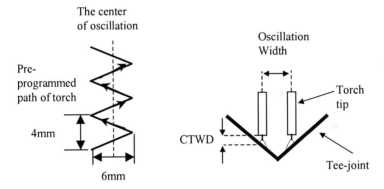

Figure 3 Sensing torch position and CTWD variation.

In order to derive a signal for the control of side or lateral deviation, information about the cross section of the grove is required. This can be done in two ways:

- The torch is oscillated periodically over the groove and the welding current is monitored. If the groove has definite flanks, the contact-tip-to-workpiece distance will change depending on the position of the torch over the groove. The distance will reach its maximum over the centre of the groove, resulting in a minimum of welding current. The signal for side correction is derived by detecting the point of minimum welding current and correlating it to the centre of oscillation.

- The torch is oscillated periodically over the groove and the welding current is measured at the extreme points of the oscillation movement. As long as the torch position is symmetrical over the groove, the contact-tip-to-workpiece distance, and thus the welding current will be the same on both the flanks. Hence,

$$I_{left} - I_{right} = 0 \qquad\qquad (2)$$

If the torch position is out of centre, a difference in the contact-tip-to-workpiece distances will be observed and a difference in the welding currents is detected and

$$I_{left} - I_{right} \neq 0 \qquad\qquad (3)$$

The amount of this current represents the distance to the centre of the groove, and the sign gives the direction of correction.

The oscillation of the arc can be achieved by a mechanical means (a separate oscillator or the welding robot) or by magnetic means [5].

A simple control algorithm [7] for reliably correcting the electrode's position relative to a vee-joint uses (i) the peak values of the current at each extreme of the torch oscillation, and (ii) the value of the current at the centre of the oscillation. The difference between the peak current values at the left and right sidewalls determines the direction and the amount of lateral correction to be applied during the next oscillation cycle. A modification to the abovementioned peak-current measurement approach is used to simultaneously centre the electrode and also control the width of the bead. This involves moving the electrode toward a sidewall until the current reaches the preset value corresponding to the correct half-width of the joint. The electrode is then directed towards the other sidewall and the process is repeated. This technique has the advantage of directly sensing the local environment at the torch-tip, which allows automatic control of the bead placement, bead geometry, and fusion. However, this also implies that the only available information about the seam is local and the equally important global knowledge (such as height mismatch, root gap, presence of tacks or previously deposited material along the joint, etc.) is unavailable.

It should be noted that the through-the-arc sensing is not easily and economically implemented in all types of welding processes and weld joints. It is only suitable for vee-groove joints or tee-joints where oscillation is needed. Butt joints and shallow seams are difficult to sense using this technique. The operation of this system depends on measuring relatively small changes in the steady-state operating condition resulting from the change in CTWD. The rate at which the system re-establishes the new steady-state operating point is primarily a function of (i) the power source characteristics, (ii) the arc characteristics, (iii) the electrode extension, (iv) the wire size, and (v) the properties of the wire material. Unfortunately, for most gas metal arc welding (GMAW) and flux-cored arc welding (FCAW), the desirable welding conditions make the time constant of the self-regulating process shorter than the torch's period of oscillations. Hence, recording the transient arc current and arc voltage requires high sampling frequencies and sophisticated control algorithms. This problem is not so severe in the submerged arc welding (SAW)) process because the large electrode size results in lower current density and therefore a longer time constant for the self-regulating process.

The through-the-arc sensing system for seam tracking is implemented in many automatic welding machines and robots. It is used for the compensation of minor tolerances arising from inaccurate workpiece preparation, inaccurate workpiece placement and thermal distortion. As no additional devices are needed near the welding gun, there are no additional restrictions concerning accessibility to the seam. As the arc sensor requires a welding process for its function it is not suited for detection of the seam start. For that reason, it is often combined with the tactile gas nozzle sensor.

There are also other sensing systems that make use of the through-the-arc-sensing principle. Systems capable of automatic adaptation of torch orientation along 2D or 3D-curved seams, allowing the welding of complex seam paths without having to program the weld path have been developed.

One system monitors the ratio of the welding current on the flanks for detection of outside corners. If a certain drop in the welding current on the upper flank occurs, an outside corner is recognised. Inside corners are detected by a gas nozzle sensor. Welding between the corners is guided by a standard through-the-arc-sensor. On reaching a corner, the welding process is stopped, the torch orientation adjusted and the welding process started again.

For tracking planar seams using the through-the-arc-sensing technique, the use of two electrodes is recommended. However, for tracking seams with three-dimensional variations, three or more electrodes are required [8].

## 2.2    System Overview

In the following sections, we will describe the development of a through-the-arc-sensing based seam tracking controller. More details on the development project can be found in [9,13].

In the case of seam tracking, (as explained in Figure 2 and Figure 3), the through-the-arc-sensing is implemented by weaving (oscillating) the torch across the joint while also moving forward along a straight line path parallel to the axis of the weld joint. The difference between the peak current values at the left and right sidewalls determines the direction and the amount of lateral correction to be applied during the next oscillation cycle. The overall control strategy is shown in Figure 4.

Figure 4 shows the flow of signals among the different subsystems. Initially, the current is sensed by a Hall-Effect sensor, which is connected to the Visual Designer board. The current signal is acquired under direct memory access (DMA). The Visual Designer board incorporates filtering and a proportional-integral (PI) controller which computes the amount of compensation required to position the torch in a lateral deviation to the seam path. The compensation signal eventually reaches the NextMove card through the data acquisition board and the host computer/controller. The NextMove card then positions the torch through a 3-axis gantry robot. Figure 4 also shows the flow of signal for starting the devices viz., fume extraction, wire feeder, and DMA, in synchronisation with the 'start weld' signal initiated by the user through the user interface.

The control aspect of Figure 4 is captured in Figure 5. The data acquisition and the main algorithm for the system control actions are carried out on a system using Visual Designer, (denoted as top-level system in Figure 5). The robot motion control is carried out on another system using Visual Basic (denoted as low-level system in Figure 5). The low-level system allows the robot to be programmed for a certain fixed-path motion; in this case, the robot is programmed to oscillate about the seam with a certain width of oscillation, as shown in Figure 3.

From the through-the-arc-sensing technique, the top-level system determines the required correction to be performed in order to maintain the oscillation at the centre of the seam. The control action output to the low-level system is in terms of a certain offset from the centre of oscillation; the low-level system thus shifts its centre of oscillation by the amount in the next cycle of oscillation.

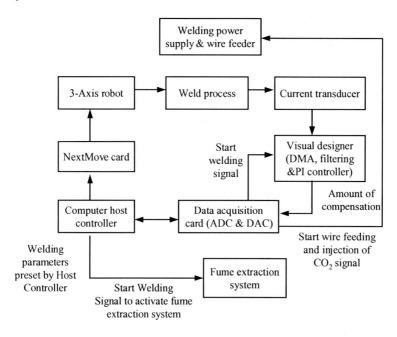

Figure 4 System overview of the seam tracking controller.

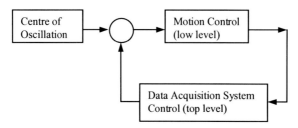

Figure 5 Block diagram of control strategy.

## 2.3    *Data Acquisition*

The top-level system is responsible for reading the arc current, and to interpret the location of the torch with respect to the seam, and output the necessary control action to bring the centre of oscillation back to the seam. The program is developed using the Visual Designer package by Intelligent Instrumentation [10]. The package allows rapid development of the entire system by making use of numerous application blocks. The simplified block diagram in Figure 6 gives an overview of the top-level system.

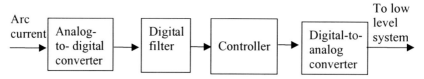

Figure 6 Simplified block diagram of the top-level system.

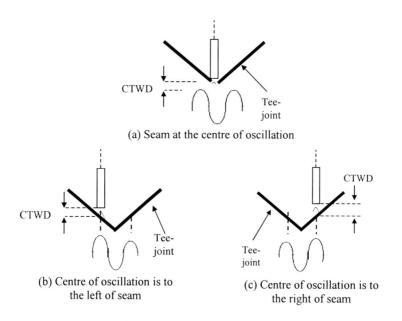

(a) Seam at the centre of oscillation

(b) Centre of oscillation is to the left of seam

(c) Centre of oscillation is to the right of seam

Figure 7 Arc current during oscillation.

Visual Designer includes both the analogue input and analogue output blocks that allow the program to input/output analogue voltages. The analogue input block is set to sample the arc-current at 500 Hz, and the data is acquired using direct memory access (DMA) transfer. Since the welding torch is set to oscillate at 1 Hz, it travels from left to right in 0.5 seconds and right to left again in 0.5 seconds, as shown in Figure 7. Hence, for every frame (0.5 second block) of data, we can obtain a peak value that corresponds to the arc current at an extreme of the oscillation. The algorithm is then set to interpret the difference between the acquired peaks in each pass to determine the control action. The control signal is output through the analogue output block to the low-level system. The maximum range of analogue voltage is +/- 10V.

## 2.4    *Signal Processing*

Figure 8 shows the signal obtained directly from the current sensor. The noise-like waveform is the result of noise introduced by the welding process which had "buried" the 2 Hz signal corresponding to the electrical oscillation frequency. In order to recover the useful waveform, a digital filter is introduced to extract the embedded signal. Since the noise is expected to be of a much higher frequency with respect to the desired signal, and the DC value of the signal has to be retained, a low pass filter with a cut-off frequency of 2 Hz is used.

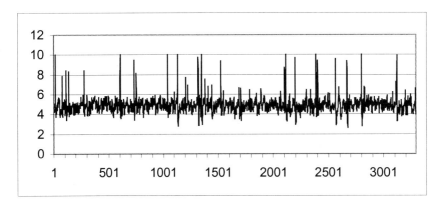

Figure 8 Unfiltered Data.

Since the amplitude of the filtered waveform is critical, a Butterworth filter with no ripples in both the passband and the stopband is chosen. The filter is configured with an order of 10, with the passband at 2 Hz and the stopband at 3 Hz [10]. The filtered signal is as shown in Figure 9.

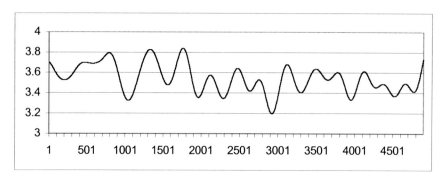

Figure 9 Filtered Data.

This stage is followed by a peak detector for two successive peaks followed by further filtering to remove low frequency noise at frequencies lower than the electrical oscillation frequency. The latter noise arises due to variations in the minute mismatch between the timings of the electrical and mechanical oscillations which gives a beat frequency effect at a relatively low frequency. Since the thermal distortion is largely DC in nature, a second filter is chosen as a low pass filter to remove the noise signals, while at the same time allowing thermal distortion effect to be monitored by the through-the-arc current. Finally, a moving average filter, at the output of the second filter, produces a signal that is sufficiently fast in response, is free from the noise signals affecting the original peak signals, and is thus suitable for control.

## 2.5    *Robotic Welding System*

The robotic welding system, used for the development project, consists of a Hobart GMAW welding DC power source, $CO_2$ gas source, three-axis robot to carry the welding torch, data acquisition system, and a motion controller card to drive the motors of the 3-axis welding robot (see Figure 10). The welding current range is up to 300 amps. The motors driving the 3-axis Robot are stepper motors. An eight-axis NextMove motion controller card was chosen, bundled with a motion control library MINT. Visual basic was used to program the motion control card, and Visual Designer to program the data acquisition card. The interface between various modules is shown in Figure 11.

Figure 10 Robotic system overview.

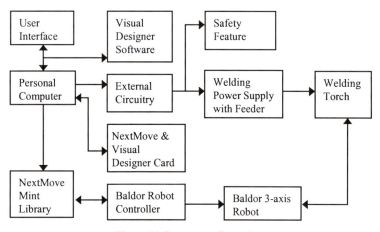

Figure 11 System configuration.

The robot was programmed to follow straight-line path of a length up to 300 mm in a three-dimensional space. This allowed for any error in fixture design to be accounted for by the robot system. The starting and the ending points of the welding seam were initially taught to the robotic system. The software automatically calculates the weaving path (of pre-determined oscillation width and pitch) from the initial to the ending seam point to be superimposed on the straight-line trajectory.

The PC used in the welding system is required to interface with many devices such as the current sensor, buzzer, emergency stop button etc. To interface these devices, I/O features, provided by the NextMove card and Visual designer card, are used. Both these cards allow analogue and digital I/O interfacing.

NextMove servo controller, for the ISA PC/AT bus, aims at motion control applications for both open and closed-loop system. It also enables the PC to communicate with the external devices by providing I/O interfacing ports. It consists of 24 digital inputs and 12 digital outputs, these ports being opto-isolated PNP or NPN and are short-circuit protected. Four 12-bits, analogue differential input ports (+/-10 V) are provided for analogue signals and are software switchable to 8 single-ended inputs ports (0 - 5 V).

The Visual Designer card, whose analogue input port is configured to be DMA analogue input port, performs continuous sampling of the feedback signal from the current sensor and also outputs a signal to a output port. The signal from the output port requires filtering to further reduce the noise within the signal before being processed and passed to the proportional-integral-derivative (PID) controller.

## 2.6    *Implementation of Seam Tracking Controller*

The basic function of the controller is to output the control signal based on the difference between two consecutive peaks. A PI controller with tenable P and I parameters is used.

### 2.6.1    *The Algorithm*

A comparison based on the difference between two consecutive current peaks is used in the controller. Since the arc current is inversely proportional to the CTWD, the difference between the left and right peaks will give an indication of the offset of the centre of oscillation towards a particular side. An ideal situation would be where the left peak equals right peak indicating that the centre of oscillation is at the middle of the seam. If the left peak value is higher than the value of right peak, the centre of oscillation has shifted to the left, whereas if the right peak is higher than the left peak, then the centre of oscillation has shifted to the right (see Figure 7). The protocol between the high-level and low-level systems is fixed such that a negative value for the compensation signal indicates a correction to the right while a positive value indicates correction to the left, the amount is directly proportional to the magnitude of the signal.

## 2.6.2 The PID Controller

The setpoint of the PID controller is set at zero (see Figure 5), corresponding to the centre of oscillation, since the control objective is to achieve the same magnitude for the left and right peak values resulting in a value of zero for the difference between two current peaks. The main part of the development was devoted to tuning of the controller's PID parameters. These gains were determined based on an initial open-loop test followed by on-line tuning.

The comparison of the arc current, at the extremes of torch oscillation, is used as the sensing strategy. This method of detecting the seam is more robust than seeking the minimum current to identify the seam. The latter method requires a noise-free arc current signal for correct operation. In reality, the arc current is very noisy. Detecting the seam using the minimum current method is not reliable. In the former method, subtracting the two current levels at the right and left extremes of the oscillation means that some of the common noise effects, affecting the two peak currents, are factored out. Moreover, such a closed-loop control system allows a stable reference of zero (the two left and right peak currents must be equal for ideal seam following). This is in contrast to other system as in [11] where the torch is adjusted to follow the minimum current offset in the open-loop without any stable reference to follow. The latter technique can only result in ideal correction, if the current signal is reliable. Noise is a common problem in arc welding, and this technique will lead to spurious correction of the welding torch.

## 2.7 Plant Identification and Control

Open-loop tests were performed on the welding process to establish a model that can be used to tune the PID controller. The derivative action was not used since the current measurement was noisy. The oscillation data, in terms of the width and the pitch of oscillation, had to be adjusted until the left and right peaks were distinctly detectable. A closer weaving of the welding torch resulted in distortion of current peaks due to previously deposited metal being detected by the current sensor. A typical width of oscillation of 6 mm and a pitch of oscillation of 4 mm (see Figure 3) were seen to give the best results. Needless to add that the other process parameters, such as, wire feed rate, voltage, and the inert gas flow rate have to be adjusted for optimal operation during welding.

Open-loop tests were conducted by injecting a pulse of -2.5 volts at the controller output and carrying out welding on a given piece. The response of the robotic welding system at the output of the moving average filter is monitored and plotted. From the open loop response so obtained, one can easily calculate the first order plus dead time (time delay) response of the plant using standard techniques [12]. The pulse direction is reversed and the open loop tests were repeated.

Tuning formulae are then used to calculate the optimal tuning parameters for the PI controller [12] given by:

$$m = K_p \left[ e + ( 1 / T_i ) \int e \, dt \right] \tag{4}$$

where $m$ is the controller output, $e$ (error) is the controller input, $K_p$ is the proportional gain and $T_i$ is the integral time.

The optimal settings were then set on the PI controller. The plant is put in a closed loop and the welding is carried out to verify the closed-loop system performance under a lateral disturbance, which was injected to the torch. Using the results, the PI controller settings were fine-tuned until satisfactory performance in the closed loop was obtained under disturbance.

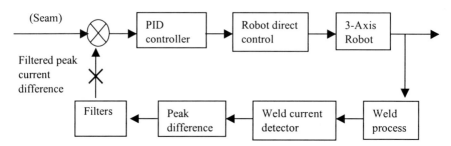

Figure 12 Control system for seam tracking using through-the-arc-sensing.

Figure 12 shows the block diagram of the control system used. The output of the system is the torch position which has a reference 'zero' corresponding to the seam position at any given time along the programmed welding path. An open loop test is performed by opening the loop at the point marked 'X' in Figure 12. A test signal of about $\Delta p = 2.5$ V is injected at the output of the PI controller. The result is shown in Figure 13.

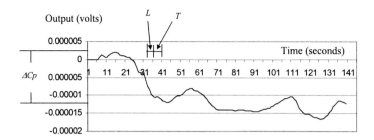

Figure 13 Open-loop process reaction curve test.

From Figure 13, using the process reaction curve method, one can calculate:

$L = 3$ s, $T = 4$ s, $\Delta Cp = 1.2 \times 10^{-5}$ volts
$N = \Delta Cp / T = 3 \times 10^{-6}$ volts/s
$K_p = 0.9 \times \Delta p/(NL) = 2.5 \times 10^5$
$1 / T_i = 1 / (3.33L) = 0.1$

where $L$ is the time delay in seconds, and $T$ the plant first-order time constant in seconds, and $\Delta Cp$ the plant open-loop steady state response in volts to a step input of 2.5 volts.

It was found that a scale factor of $10^5$, and $K_p = 1.0$, $T_i = 0.1$ provided the best performance for the controller. Figure 14(a) shows the result of open loop test with a pulse injected at position 'X'. The weld can be seen to be veered off from the seam. Figure 14(b) shows the result of pulse test under the closed loop. It can be seen that the effect of the pulse has been corrected and the seam is followed by the weld.

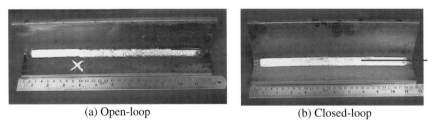

(a) Open-loop                                          (b) Closed-loop

Figure 14 Open-loop and closed-loop comparison with pulse injection.

Another test involved was to offset one of the end points off the seam by 5 mm. In the open loop, the torch followed the pre-programmed path without making any correction resulting in a weld, which is away from the actual seam, as demonstrated in Figure 15(a). Under the closed loop, the PI controller corrected the torch position during welding continuously, resulting in a weld which is exactly on the seam as can be seen from Figure 15(b).

(a) Open-loop (close-up view)

(b) Closed-loop (close-up view)

Figure 15 Open-loop and closed-loop comparison (ramp input).

Figure 16 Angle weld.

Figure 15 shows the actual results on a 280 mm long fillet-joint. We tested our controller on a piece of angle weld by teaching only start and end points. The result showed that it followed the seam instead of joining these two points (see Figure 16).The results obtained through welding under the closed loop, as described above, are representative of many such trials conducted to verify the closed-loop design and implementation.

## 3.    Vision-Based Seam Tracking

### 3.1    *The Technology*

A vision based seam tracking controller consists of the following:

*   A structured light based camera.
*   Camera control unit.
*   Image processing module.
*   Robot with controller.

The sensor operates on the principle of active triangulation ranging. A sheet of light is generated using a combination of a laser diode and a cylindrical lens. This sheet of light is arranged to fall across a camera field of view at a known angle. The image resulting from the intersection of this light sheet with the workpiece is in the form of two curved or straight-line stripe segments corresponding to the two pieces of metal to be joined. The sensor is located immediately in front of the welding torch. Each strip position in the image maps to a corresponding 3D position of the joint, which can be found by analysing the strip sections. The tracking error can be determined by comparing this measured joint position with some taught position. Both lateral (side-to-side) errors and stand-off addition to these positional errors can be detected. The strip analysis also allows the seam gap or fit-up error to be determined.

The problems of observing the image resulting from the light strip is complicated by a number of factors, most particularly the high level of spurious information in the image. The most commonly observed spurious information may be characterised in terms of arc glare, weld spatter and specular reflection.

The light radiated from the arc and the molten weld pool beneath is very intense. The emission covers a broad range of wavelength, with the peak energy density in the ultraviolet (UV) region. The wavelength sensitivity of the camera is very broad, extending from the near UV to beyond 1000 nm in the infrared (IR). The laser output is, however, virtually monochromatic at a wavelength in the region of 830 nm (near IR). This allows the use of a very narrow band interference filter (half-power bandwidth is approximately 8 nm) tuned to the laser's wavelength in order to reduce the intensity of the unwanted background illumination by at least two orders of magnitude. Further reductions, can be achieved with a suitable mechanical design of the camera. Nevertheless, the arc glare is not completely removed and it may still be brighter than the stripe in parts of image.

As a side effect of GMAW process, small beads of molten metal (typically 1 – 2 mm in diameter) are ejected from the weld pool. Some of the particles may fly across the field of view of the camera. Since they are red hot they radiate a significant amount of energy in the near IR and are therefore visible to the camera/filter combination. Traces of such particles appear in the majority of the images and they must be removed. As they travel above the object plane they produce out-of-focus images. In the charge coupled device (CCD) imaging, the effect of the fast passage of a luminous object in this manner is the addition of a 'trail' of increased brightness to the original image. Since the camera field of view lies ahead of the weld pool along the seam direction, and the spatter emanates approximately radially from the weld pool, these trails are approximately parallel to the seam direction, in contrast to the stripe, which is within 25° of the perpendicular to the seam direction. In addition, trails tend to be much broader than the stripe, and this aids discrimination between the two.

The biggest problem results from the nature of the metal surfaces, which act more as specular reflectors than as diffuse scatterers. The resulting image varies considerably in intensity depending upon the angles of the joint. The tee-joint configuration can act as a corner reflector, giving rise to a strong reflection of the stripe from each metal surface on the other. This spurious reflected stripe is often more intense than the original projected stripe, but it is badly out of focus and therefore much wider than the original.

The simple image thresholding techniques are ineffective in addressing the above problems. A spatial filter that operates in a direction parallel to the seam is the solution.  A one-dimensional linear Difference-Of-Gaussians (DOG) filter provides satisfactory results in practice and it has the advantage that it can be easily implemented in hardware.

## 3.1.1 Triangulation Techniques

Three types of optical scanning technology are available for the imaging of a 3D object.

A point imaging sensor is similar to a physical probe in that it uses a single point of reference, repeated many times. This is the slowest approach as it involves lots of physical movement by the sensor.

The area imaging sensor is technically difficult, demonstrated by the lack of robust area systems on sale.

Stripe imaging systems have been found to be faster than point probing as they use a band of many points to pass over the object at once. They are accurate, thus matching the twin demands for speed and precision.

Most of the techniques for absolute 3D measurement are based on triangulation, i.e. calculation of the absolute distance from a uniquely defined triangle. The triangulation is called active, if the corners of the triangle are formed by the point that has to be measured, the light source and the sensor. It is called passive, if the active light source is replaced by another sensor, which gives two sensors and the point that has to be measured as the corners of the triangle.

## 3.1.2 Measurement Range and Accuracy

The lateral measurement of active techniques, i.e. the range in $x$ (along the seam path) and $y$ (across the seam) directions, can range from millimetres to meters in size. The limitations are set by the intensity of the light source. The range in $z$ (vertical) direction is about 50% of the lateral size.

The lateral resolution depends on the image field and on the resolution of the optical sensor. This resolution is for standard CCD cameras, and for special CCD cameras. Some object specific data can be calculated with so called "sub pixel algorithms" with an accuracy of 1/10 pixels, which means a lateral resolution of 1/5120 with a standard CCD camera.

The resolution in $z$ direction depends on the image field and on the angle of triangulation, i.e. the angle between the projector and the camera axis: A larger angle means a higher resolution but also more shadows and less scan data. The accuracy in the $z$ direction is 1/4000 of the image field size.

### 3.1.3    The Principle of Laser Triangulation

A point of light is projected from the laser diode to the object being measured. The light scattered from the object is imaged onto a light-sensitive detector. As the distance from the sensor to the surface changes by $\Delta z$, the light reflected on the surface is imaged to a new position on the detector, say $\Delta x$. This position on the detector $\Delta x$ can then be correlated to an accurate $\Delta z$ measurement.

## 3.2    Vision Based Seam Tracking Systems

Machine vision was used in the early 1980's for the off-line determination of workpieces to be welded. Seam tracking systems were developed for the correction of single-pass robot paths. The measurement and correction of weld pool geometry for welding process control was also attempted. However, most early vision based robotic welding systems did not find wider acceptance in industries because of limited functionality and flexibility. Their applications were also restricted to one application area or even to a single joint configuration. These systems that can be classified as first generation vision systems suffer from the following disadvantages:

- Sensitivity to torch orientations.
- Sensitivity to surface finish.
- Sensitivity to welding smoke.
- Sensitivity to spatter (or flying hot metal pieces).
- Inability to deal with tack welds or previously deposited material.
- Inability to compensate for joint shape variations by controlling the welding process conditions (adaptive welding).

Much research was carried out in this area to improve vision sensors and welding process control schemes for reliable and wider application of robotic welding. This has led to the development and implementation of flexible second-generation systems for vision-guided seam tracking and adaptive welding. Structured lighting approach is used for:

- The off-line detection of large fixturing errors before welding starts.
- Real-time correction of robot paths to compensate for thermal distortion and loose part tolerances.

- The in-process adjustment of welding conditions to correct for weld joint shape variations.

Advanced hardware and software with multi-processing capabilities is required for these applications. Special software programs are necessary to perform various tasks which include: robot motion trajectory planning, multi-axis servo control, image acquisition, image processing, 3D workpiece surface modelling and robotic welding process control based on sensor feedback.

## 3.2.1   System Architecture

The overall system architecture of a vision system for weld seam tracking is shown in Figure 17.

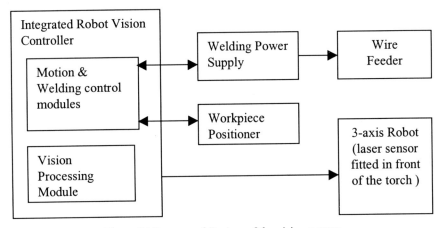

Figure 17 System architecture of the vision system.

The components of this vision based tracking system include the following:

Hardware:

- Laser camera with controller unit.
- Industrial PC.
- Robot.
- PC based motion controller card.
- PC based frame grabber card.

- GMAW system.

Software:

- Dedicated vision processing algorithms
- Seam tracking & control algorithms for robot

We review, briefly, some of the commercial systems available.

### 3.2.2    Vista Weld Laser Camera

The Vista Weld sensor can be used for welding applications to determine the exact position of the seam before and during welding. A product of Integrated Sensor Systems Ltd, UK, the laser sensor can be integrated with most automatic welding equipment including circumferential and linear machines.

The VistaWeld sensor projects a stripe of visible laser light onto the weld joint. A built-in video camera images the reflected laser light and transmits the video signal over an armoured fibre optic cable to a frame grabber board in the PC. The VistaWeld sensor includes an optical filter which ensures that the sensor signal is unaffected by the light from the welding arc.

### 3.2.3    Two Beam Laser Camera

The laser sensor  uses two 30 mW semiconductor laser diodes and special optics to project the point beam as line beam. The two laser beams overlap giving a wider range. The device also houses a tiny CCD camera to image the reflected beam from the workpiece. The camera outputs a video signal, which can be digitised and processed to get the seam points during welding. The camera has been designed to be used in welding applications and has provision for cooling the camera using either compressed air or water. This is a product from Meta Technology, UK which specialises in sensors and automation.

### 3.2.4    Laser 3D Vision Sensor

A laser 3D camera for welding application was developed by Servo Robot, Inc. The camera has been integrated with many welding robots. The vision system consists of the sensor, vision software modules and special vision hardware controller. A variety of laser sensors are available for different applications with various depths of field. The system is claimed to detect edges with an accuracy of about 0.5 mm. The camera, with the standard software, is able to detect shapes of joints of any type, vee-grooves, fillet-grooves, lap joint, butt joint etc. The information obtained from the optical signal, allows one to perform the following:

- Detection and measuring of gap.
- Calculation of position of centre line of joint.
- 3-D seam tracking.
- Automatic volume compensation.
- Inspection of welding system.

### 3.2.5    Point Laser Scanner

A seam tracking system developed by the Industrial Robot Division of ASEA AB uses a laser sensor to position and track the seam in real time. The main application is for welding of car bodies where sheet metal is used.

The seam finding system consists of an optical sensor and an integrated microcomputer which evaluates the sensor signals and transmits the result to the adaptive functions in the robot control system. The sensor is mounted on the torch holder and the measuring spot is 20 mm from the wire tip. The sensor head weighs 650 grams and is environmentally protected against heat, fumes and spatter by flow of clean air and two shielding glasses.

### 3.2.6    Laser Camera

A modular seam tracking system was developed by the University of Waterloo. It provides real-time control of the robot trajectory based on a vision sensor (laser camera) data and weld process feedback. The system, independent of robot type or manufacturer, provides a "plug-in" solution and allows a variety of control schemes to be implemented. It enables a workpiece positioner to be fully integrated with the robot. Control of the robot and the workpiece positioner is entirely based on real-time seam sensing data. This enables the robot system to 'teach itself' the seam trajectory, thus greatly simplifying the programming effort.

A full frame sensor was used to get the cross-section of the workpiece surface below the sensor. Figure 18 shows the weld profile obtained for a fillet geometry. Each image is analysed to obtain the location of the seam as well as the root gap dimension.

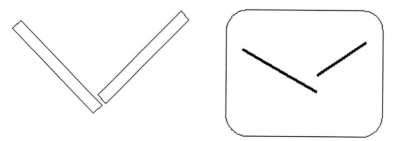

Figure 18 Workpiece geometry and sensor image.

The distinctive feature of dynamic seam tracking is that it runs independently of the robot controller. After the robot is taught a path, a command cycle is entered. It enables the intelligent sensor interface to provide correction to a pre-programmed trajectory, or completely control the motion of the end-point while tracking the seam. The robot program is "unaware" of its seam tracking capability.

### 3.2.7    Fanuc's MIG EYE

To automate the GMAW process, Fanuc developed a sensor unit called MIG EYE for seam tracking applications. The sensor unit is mounted on the robot wrist and a laser beam is projected on the surface of the job. The position of a welding joint is calculated from the data of the beam received. A visible laser, 690 nm in wavelength, is employed to detect welding joint. The sensor not only tracks the seam, but also senses the variation in gap and changes welding parameters to achieve good quality consistently.

The sensor unit consists of a laser module, CCD module, the galvanometer module, and the casing. The laser beam, generated by the laser module, is deflected by the galvanometer module and emitted from the sensor. The CCD module detects the laser beam being reflected from or diffused by the workpiece. The position of the weld joint can be calculated from the detected position on the CCD module and the angle of the mirror in the galvanometer. A pre-unit processes the signal from the CCD module, and feeds the resultant joint data to the MIG EYE CPU for further processing.

## 3.3    *System Description*

The objective of the development reported here was to design and develop a vision based seam tracking system suitable for application in welding using robotic systems. The following pages will describe the development of such a system called vision-based seam tracking (VisTrack).

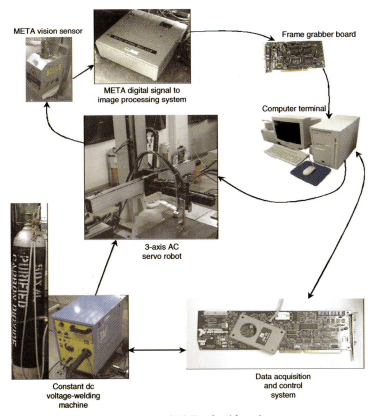

Figure 19 System layout of VisTrack with major components.

## 3.3.1    *Hardware-System Layout*

The hardware used consists of a META vision sensor, a META digital signal to image processing system, a frame grabber board, a computer terminal, a 3-axis AC servo robot, and a constant DC voltage-welding machine together with a data acquisition and control system. The basic configuration of the whole system layout is shown in Figure 19.

### 3.3.2    Sensor Hardware

The main components of the sensor head consist of a CCD camera, two laser diodes and a microprocessor as shown in Figure 20. The two laser diodes project a stripe of laser beam onto the weld joint. A CCD camera, through a prism, band-pass optical filters and lens unit, views the reflected laser beam. The components of the sensor head mentioned above are vulnerable if left unprotected.

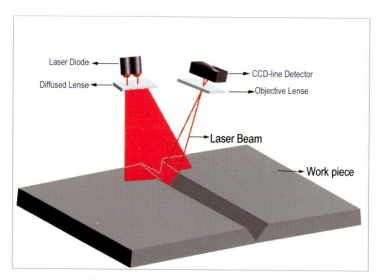

Figure 20 Main components of sensor head.

During the welding process, the temperature of the workpiece increases and so does that of its surroundings. In order to maintain the working temperature of the sensor, chilled air at slightly above the atmospheric pressure is led through the sensor in special channels (as shown at the top of Figure 21) to reach all of the interior parts.

The air is forced out through the light openings and thereby hindering spatter and smoke from reaching the optics inside the sensor head. In addition to the chilled air, an optional water jacket can be used for further cooling.

At the opening at the bottom of the sensor head (see Figure 21), there is a replaceable protective glass window to act as a barrier for spatter that may arise from the welding. After numerous welding, spatters may soon cause stains to form on this protective glass. These stains will eventually affect the quality of the images captured, and hence it is necessary to replace this glass.

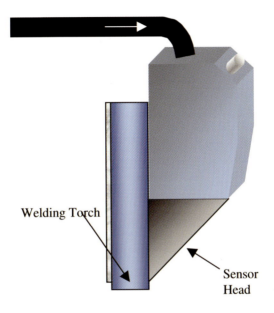

Figure 21 Sensor head.

### 3.3.3    Mounting of Sensor Head

The sensor head, which includes the laser diode and the CCD camera, is mounted in front of the torch, as shown in Figure 22. Also included is the air supply used for cooling the interior of the sensor head.

With this mounting, the laser diode in the sensor head will project a laser stripe across the workpiece in front of the torch. The horizontal distance between the torch's tip and the laser stripe is known as the look-ahead distance, while the vertical distance between the sensor head and the workpiece is known as the stand-off distance. The definition of the two distances is shown in Figure 23.

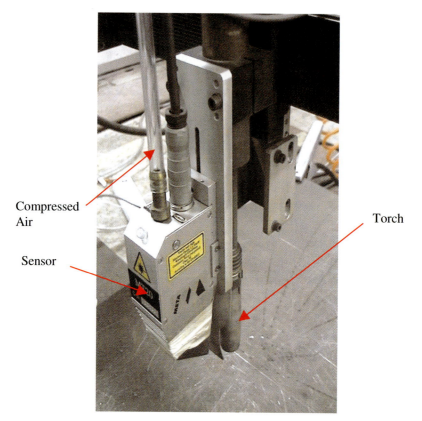

Compressed
Air

Sensor

Torch

Figure 22 Mounting of sensor head.

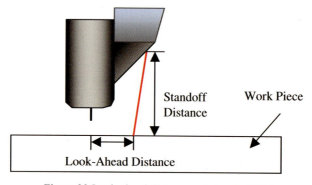

Standoff
Distance

Work Piece

Look-Ahead Distance

Figure 23 Look-ahead distance and stand-off distance.

### 3.3.4    Data Acquisition System

The data acquisition system consists of a current transducer, a voltage transducer, and a PCI data acquisition card from National Instruments, as shown in Figure 24.

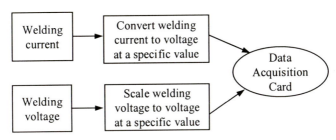

Figure 24 Data capturing.

The current transducer converts the welding current (0 – 400 A) to a voltage signal (0 – 10 V). As the welding current is quite high, it is not possible to measure it directly. The current transducer uses the Hall-Effect method to sense the welding current. The voltage transducer uses the voltage divider method to scale down the welding voltage by 4 times, and converts the voltage from 0 – 40 V to 0 – 10 V.

Figure 25 PCI data acquisition card.

The PCI data acquisition card, shown in Figure 25, plugs directly to the PCI bus of the system computer, which greatly extends the data throughput rate up to 132 Mbytes/sec, and also has provisions for processor-free direct memory access. During the bus mastering, the PCI DAQ board takes control of the PCI bus, transfers data at high rates of speed, and then releases the bus for other peripheral use.

Two analogue input channels are used to acquire the welding current and voltage, and two digital output channels are also used. One of the digital output channels is used to switch on/off the welding machine. Another channel is used as a link between the main program and the data acquisition program.

### 3.3.5    Software Overview

Figure 26 shows how the VisTrack system is interfaced to external devices. Table 1 lists the main software modules in the VisTrack control system.

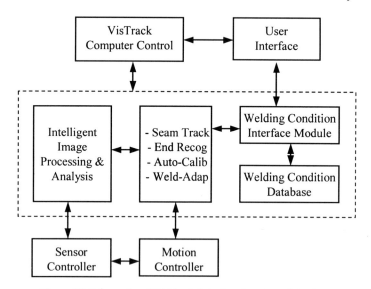

Figure 26 Schematic of VisTrack interface to external devices.

The entire seam tracking program was written using Microsoft Visual Basic 6.0 working under the Windows NT 4.0 Operating System environment, including the user interface, the VisionBlox image processing tools and other image processing techniques. The movement of the 3-axis robot-welding torch is controlled using MINT Interface Library codes written in Visual Basic, via the I/O interface card (NextMove Card). The current and voltage data acquisition is also controlled using functions written in Visual Basic, via the I/O interface card (from National Instruments). These are summarised in Figure 27.

Table 1 VisTrack software overview.

|  | **Description** |
|---|---|
| **Standard** | • Seam Tracking<br>• Automation Calibration Software<br>• External (User) Interface Unit |
| **Option** | • Depth Sensing & Adaptive Welding Process Control<br>• Recognition of Welding End Point<br>• ROI Auto-Placement<br>• Simple Pre-Weld Set-up<br>• Real-time Capturing and Logging of Welding Parameters<br>• Historical View of Vital Data for Post Weld Analysis |

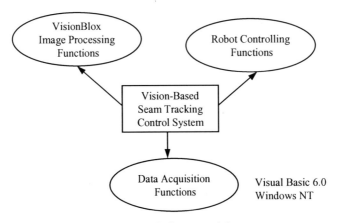

Figure 27 Software modules.

## 3.4    *Image Acquisition and Processing*

### 3.4.1    *Structured Light Approach*

The structured light sensor operates on the principle of triangulation as shown in Figure 20. A sheet of light is generated using a combination of a laser diode and cylindrical lens. This sheet of light is arranged to fall across the camera field of view at a known angle. The detector (CCD camera or CCD-line detector) receives the reflected laser beam at an angle. An optical filter (band pass filter) will filter off all other light rays arising from welding, only allowing light at the same frequency as that of the laser beam. At different depths, the detector receives the reflected laser beam at different angles and the depth can then be calculated by triangulation.

Although the camera is mounted above the workpiece, it is able to detect the depth changes along the cross-section of the workpiece, i.e. the image captured is similar to looking at the workpiece from the front view. The image is then passed through a hardware filter to remove noise. It is then digitised and stored in pixel memory by a frame grabber board.

### 3.4.2    Capturing the Image

VisionBlox image acquisition and preparation tools are used to grab an image. The programming of capturing an image is summarised in Figure 28.

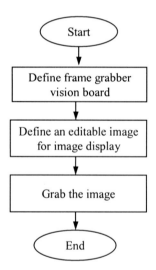

Figure 28 Image acquisition programming.

The process of capturing the image is summarised in Figure 29. The size of the digitised image is 768 by 572 pixels with a greyscale level of 0 to 255. The origin of the image (0,0) is at the top left-hand corner of the image (see Figure 30).

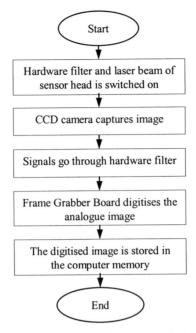

Figure 29 Processing of image acquisition.

Figure 30 A 768 by 572 pixels captured image with origin (0,0) at top-left corner.

### 3.4.3    Image Preprocessing

Before the captured image can be processed to obtain the seam point, it has to be "cleaned up" first or to undergo some image preprocessing. Image preprocessing can be carried out using built-in functions from VisionBlox.

## 3.4.4    Erosion

Erosion works (at least conceptually) by translating a structuring element to various points in the input image, and examining the intersection between the translated kernel coordinates and the input image coordinates. It takes two pieces of data as inputs (see Figure 31). The first one is the image that will be eroded. The second is a (usually small) set of coordinate points known as a structuring element (also known as a kernel). It is this structuring element that determines the precise effect of the erosion on the input image.

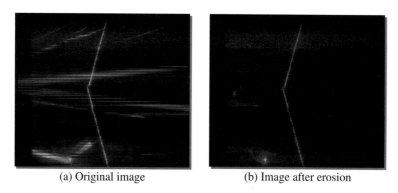

        (a) Original image                  (b) Image after erosion

Figure 31 Original image and image after erosion.

The effect of greyscale erosion will generally darken the image. Bright regions surrounded by dark regions shrink in size, and dark regions surrounded by bright regions grow in size. Small bright spots in images will disappear as they are eroded away down to the surrounding intensity value, and small dark spots will become larger spots. The effect is most marked at places in the image where the intensity changes rapidly, and regions of fairly uniform intensity will be left more or less unchanged except at their edges

In VisionBlox, erosion is used mainly to eliminate fine detail from features, enlarge gaps and holes in features, or decrease the size of a feature without changing its general shape.

## 3.4.5    Blobs Tool Control

The Blobs Tool Control is used to segment the inspection image into areas of similar intensity. Information concerning the blobs, such as size, number, and location, is then made available to the user. Blobs Tool Control performs a blob, or connectivity analysis, on the inspection image and displays the results on the inspection image (see Figure 32). This process will segment the inspection image into areas of the blob. Blobs Tool Control can operate on the whole image. A user can also define a region of interest (ROI) using an editable shape control.

Figure 32 Image after undergoing erosion, thinning and blobs tool.

## 3.5 Seam Detection Algorithms

Different algorithms have been considered for implementation of the seam tracking controller [13,18]. These include methods involving template matching, feature finding, Hough Transform and detection based on pixel intensity. We describe below seam detection algorithm using 'feature find' as it was found to be superior to the other algorithms considered for the task at hand.

The primary function of the feature-find tool is to locate a given feature within an image. The feature-find tool uses a normalised correlation engine. If a feature is located with this tool, a resultant position is made available. Each find has an associated correlation score, the angle at which the model was found, and status flags indicating that the model was found within the region of interest, or if it was found on an edge.

One reason the feature finding tool is so fast is that it uses a series of heuristics to speed up processing. Consider a normal convolution process where a certain size kernel is moved over the mage – pixel by pixel, this image matching comparison is normally a processor intensive and slow process. By using sub-samples, on both the search image and model, regions indicating high potential matches are found more quickly.

Once the feature finding tool is executed, it will indicate whether it successfully found its model by setting the "found" property to true. When the "found" property is set to true, then the $X$ and $Y$ coordinates of the found models can be read by "$X$ Position" and "$Y$ Position" property.

In this application, only one trained feature or reference feature, which is normally at the starting position, is used. Whenever the system captures a new frame, it will use this trained feature to locate any similar feature in the acquired image.

As the process is started, the system uses the trained feature to locate the seam point from every image acquired. Once the identical feature is found, the "found" property is set to true, and the $X$ and $Y$ coordinates of the identical feature are read in pixels. However, the $X$ coordinate represents the vertical axis ($z$-axis) of the real world.

### 3.6    *Implementation and Experimental Results*

A unique feature of the VisTrack system is that only a starting point needs to be taught to the robot. Once this starting position has been taught, the robot system is simply instructed to weld a specified distance. This system is based on a form of end-point control that we refer to as "Dynamic Seam Tracking". A laser-profiling sensor located ahead of the welding torch is used to detect the seam position, orientation and geometry. As long as the seam is detectable from the starting position, the system will automatically track and weld the seam for the specified distance, without any a priori knowledge of the seam trajectory.

### 3.6.1    *Evaluating Accuracy of Seam Tracking*

For the purpose of evaluating the accuracy of tracking a seam using VisTrack system, an experimental test under the following conditions was carried out:

| | |
|---|---|
| Joint type: | Vee-groove, fillet, and butt |
| Steel sheet thickness: | 2 – 20 mm |
| Type of weld line: | Straight line, tilted straight line, zigzag |
| Welding speed: | 5 – 15 mm/sec |

Examples of experimental results for a vee-groove joint, butt joint and fillet joint are shown in Figure 33 to Figure 35.

Figure 33 Zigzag vee-groove joint.

Figure 34 Straight line butt joint.

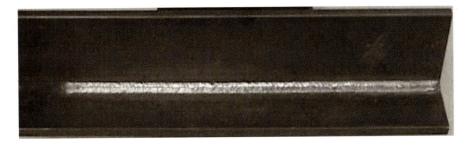

Figure 35 Straight line vee-groove joint tilted horizontally by 26 mm.

Based on the shape of the beads, it is clearly shown that the torch was moved smoothly without jerking. Further tests were performed and the following observations were obtained.

- The system can easily detect a seam position even though the workpiece is placed tilting to an angle of ±15°.

- It has the capability of sensing the depth of deviation.
- It is also suitable for many types of weld joints like vee-groove, butt, lap, fillet etc.

### 3.6.2    Evaluating the Overall Accuracy

Different experiments were conducted in order to isolate the errors due to different components, such as, the motion encoder, motor synchronisation, image processing tools etc. [13,18].

The overall accuracy of tracking a seam can be calculated using the following formula:

$$\varepsilon = \sqrt{\varepsilon_1^2 + \varepsilon_2^2 + \varepsilon_3^2}$$

where $\varepsilon_1$ = Vision Error Accuracy, $\varepsilon_2$ = Servomotor Encoder Error Accuracy and $\varepsilon_3$ = Synchronisation Error Accuracy. It is estimated that

$$\varepsilon = \sqrt{0.1^2 + 0.05^2 + 0.35^2}$$
$$= 0.367 \text{ mm}$$

We concluded that the positioning error of the torch could be corrected to less than 0.4 mm, to obtain accurate and good welds. These figures are obtainable at welding speeds up to 15 mm/sec.

## 4.    Concluding Remarks

Weld quality can be defined both in terms of the uniform and accurate tracking of the weld seam as well as obtaining appropriate dimension and shape of the welded bead itself. While the human welder can skilfully meet both the objectives under variable operating conditions, the problem is quite difficult when one tries to duplicate the human performance on a robotic welding system.

There is a need to feed several sensory inputs to the robotic system so as to derive appropriate parameters relating to weld quality and take any corrective action needed. The sensory inputs used in the project are welding current, voltage, wire speed, welding speed, and images using a laser vision system.

In this chapter, we have summarised the seam tracking techniques available using the Through-the-arc-sensing and the vision-based technologies. We have also reported on the development of cost-effective products in place of the proprietary and expensive commercial products of similar scope.

The through-the-arc-sensing controller developed can be used for certain joint types, such as fillet joint, where weaving of the torch for uniform deposition of material can be employed. The technique developed is unique in the sense that the principle used is based on a null method involving the difference of two currents. Thus any common noise is cancelled out, paving the way for the use of the technique with a commonly available welding power source which can be highly noisy.

The seam tracking system developed with a structured light vision system uses a simple algorithm based on a correlation technique. The tool needs to be initially trained for a given joint. After training, the vision system is able to guide the torch along an unknown path geometry.

It is a robust technique which can be useful for different types of joints, is less susceptible to loss of accuracy due to reflectivity of the joint surface and is relatively fast.

This technique should find applications in many areas since seam tracking is a basic requirement in almost all applications of welding automation.

## References

1.  Agapakis, J.E., Chickering, B.R., Renwick, R.J., and Wall, M.R. "Manufacturing applications of a flexible adaptive robotic welding system", Proceedings of the Flexible Welding and Cutting Conference, October 6-8, 1987.

2.  Cary, H.B. "Modern Welding Technology", Prentice Hall, 1998.

3.  Cook, G.E. "Robotic arc welding: Research in sensory feedback control", IEEE Trans. on Industrial Electronics, Vol. IE-30, No. 3, August 1983, pp. 252-268.

4.  Lane, J.D. (Editor), "Robotic Welding - International Trends in Manufacturing Technology", IFS Publications, UK and Springer Verlag, Berlin, 1987.

5.  Dilthey, U., and Stein, L. "Through-the-arc-sensing: A multipurpose low-cost sensor for arc welding automation", Welding in the World, Le Soudage Dans Le Monde, Vol.34, September 1994, pp. 165-171.

6.  Nayak, N., and Ray, A. "Intelligent Seam Tracking for Robotic Welding", London, Springer Verlag, 1993.

7.  Cook, G.E., et al, "Computer-based analysis of arc welding signals for tracking and process control", IEEE Trans. on Industrial Electronics, Vol. 34, 1986, pp. 1512-1518.

8.  Raina, Y., "Double Electrode Seam Tracking System Using Through-the-Arc Technique", Master's Thesis, The Pennsylvania State University, 1988.

9.  Wang, J.Y. " Through-the-Arc-Sensing Technique for Seam Tracking", Master of Science Dissertation, School of Electrical & Electronic Engineering, Nanyang Technological University, Singapore, 2000.

10. "Visual Designer Reference Manual", Intelligent Instrumentation Inc, 1997.

11. "IGM Robotic System, Mechanical and Electrical Manual", IGM, Austria.

12. Smith, C.A., and Corripio, A.B. "Principles and Practice of Automatic Process Control", John Wiley & Sons, 1985.

13. Devanathan, R. "An Intelligent Multisensory Knowledge-based System for Weld Quality Control", Upstream Project (Ref No: U96-A-044) Technical Report, School of Electrical & Electronic Engineering, Nanyang Technological University, Singapore and Gintic Institute of Technology, Singapore, June 2000.

14. Agapakis, J.E., Katz, J.M., Friedman, J.M., and Epstein, G.N. "Vision-aided robotic welding: An approach and a flexible implementation", The International Journal of Robotic Research, October 1990, Vol.9, No.5, pp. 17 – 34.

15. Agapakis, J.E. "Vision-Aided Remote Robotic Welding", Massachusettes Institute of Technology, Ph.D. thesis, 1985.

16. Nishie, K. "A Seam Tracking System with a Visual Sensor", 1987.

17. Tan, B.Y., and Looi, W.L. "Three Dimensional Vision System Using the Structured Light Approach", Final Year Project Report, School of Electrical & Electronic Engineering, Nanyang Technological University, 1998-1999.

18. Ching, Y.F., and Koh, K.K, "Vision-Based Seam Tracking Controller System", Final Year Project Report, School of Electrical & Electronic Engineering, Nanyang Technological University, 1999-2000.

19. Fisher, R., Perkins, S., Walker, A., and Wolfart, E. "Hypermedia Image Processing Reference", J. Wiley & Sons Ltd, 1996.

20. Vision Blox Custom Control Reference Manual, Version 2.2, Jan. 1998.

21. Chen, X.Q., Smith, J., and Lucas, J. "Microcomputer controlled arc oscillation for automated TIG welding", Journal of Microcomputer Applications (1990) 13, pp. 347-360.

# CHAPTER 6

# WELD POOL GEOMETRY SENSING AND CONTROL IN ARC WELDING

Hong Luo* and Rajagopalan Devanathan**

*Gintic Institute of Manufacturing Technology,
71 Nanyang Drive, Singapore 638075

**School of Electrical & Electronic Engineering, Nanyang Technological
University, Nanyang Avenue, Singapore 639798

## 1.    Introduction

The welding process is physically complex, non-linear and dynamic. It is very difficult to get a mathematical model to simulate and control such a process. But, the geometry of a weld pool contains information about the welding process and welding quality. Thus weld pool sensing and control play a significant role in automated arc welding.

Regarding observation of the weld pool, one problem is that the weld pool is surrounded by very bright arc emissions. Another difficulty is that, in addition to the weld face geometry, measures of weld penetration and backside weld width are also required even though they cannot be seen by topside sensors.

Among the existing methods, weld pool oscillation has been extensively studied [1-5]. An abrupt change in the oscillation frequency of the pool during the transition from partial to full penetration was found. It also has been recognised that oscillations in the weld pool could provide information about the size of the pool, which might be used for in-process control of weld pool geometry.

Ultrasound based weld penetration sensing has also been extensively investigated [6-10]. Although significant progress has been made, practical applications are still restricted because of the contact sensor. A non-contact laser phased ultrasound array is currently being used as a new solution [10].

Since the temperature distribution in the weld zone contains abundant information about the welding process, infrared sensing of welding process has been explored [11-14]. Valuable results in this area have been acquired. The penetration depth of the weld pool has been correlated with the characteristics of the infrared image.

Because of the plasma impact, the surface of an arc weld pool is depressed. It has been found that the depression of the weld pool can be correlated to the penetration depth of the weld pool [15-17]. However, the sensing of the surface depression is difficult. As an alternative, a vision system to measure the sag behind the weld pool was developed and it has been found that the average sag depression of the solidified weld bead has a good linear correlation with the backside bead width [18]. However, there is an inherent measurement delay if the weld behind the pool is monitored rather than the pool itself. A novel mechanism for observing the pool surface shape was proposed [19-20]. Laser stripes projected through a grid were reflected from and deformed by the mirror-like pool surface. The shape features of the weld pool surface were clearly shown by the reflection pattern.

As for welding process control, efforts were mainly put into penetration control. Since penetration cannot be seen by topside sensors, it needs to be inferred from available weld pool length and width through modelling. Various approaches, such as, adaptive control, model based fuzzy logic control and interval model based control, have been explored [21-25].

Although significant achievements have been made in the area of weld pool geometry sensing and control, more accurate and reliable techniques are still needed. It is understood that topside weld pool observation could provide much more critical, accurate and instantaneous information on the welding process than most of the existing methods. Due to a recent development in sensing technology, clear imaging of the weld pool is possible by a laser assisted high shutter speed camera [26].

Human welders adjust their operations based on their weld pool observation. This implies that topside weld pool geometry can be used to estimate and control the welding fusion state without knowing pool penetration, which will make the control process more amenable to real-time work. It also implies that an advanced control system could be developed to control the fusion state by emulating the estimation and decision-making process of the human operators.

In this chapter, a survey of weld pool sensing will be given. Then a vision based weld pool geometry control system using neurofuzzy logic will be discussed in detail.

## 2.     Survey of Weld Pool Inspection

### 2.1     *Weld Pool Oscillation*

### *2.1.1     Sensing Principle*

Weld pool oscillation could be triggered in a number of ways, for instance: by mechanical vibrations, by the impact of droplets entering the weld pool, by plasma arc force, by gas bubbling and by sudden changes in arc current (sudden changes in arc pressure).

The concept of using weld pool motion as a pool geometry sensing method was proposed by Hardt [1] and later demonstrated by Zacksenhouse, Richardson, Renwick and Sorensen [2-4] in the 1980's. The concept is based on the fluid dynamics of a pool constrained by a solid container and by significant surface tension forces. Such a pool will exhibit a surface motion that is a function of external forces, the properties of the fluid, the surface tension, and the shape of the container. Thus, if this motion can be excited, measured, and related to the pool geometry, a means of sensing pool shape will exist.

### *2.1.2     Theoretical Models*

The goal of the models is to obtain expressions for the dominant frequencies and modes of oscillation of a weld pool, but not to predict the complex fluid flow within the molten region. For this reason, the internal effects of electromagnetic stirring, thermal gradients, etc., are not considered. The models only predict the gross motion of the molten region based upon boundary conditions imposed by the geometric constraints.

### *a. Model One for Stationary and Moving Pool*

The welds to be studied are full penetration welds on a thin steel plate. To obtain a simple model, the molten pool is assumed to be suspended by surface tension from the surrounding solid weldment. The stationary weld is modelled as having a circular surface and the thickness of the molten region is also assumed to be small compared to its diameter. The pool can thus be modelled as a thin membrane that satisfies the wave equation. The resulting first and second mode frequencies are given by:

$$f_0 \approx \frac{156}{w} Hz \quad f_1 \approx \frac{249}{w} Hz \tag{1}$$

where $w$ is the weld pool width in mm and $f_0$ and $f_1$ are the oscillation frequencies in Hz.

A similar analysis can be made for a moving pool. To observe the effect of the altered pool shape on its natural frequencies, the pool is modelled as having an elliptical shape with an eccentricity $\varepsilon$ given by

$$\varepsilon = \left[ 1 - \left( \frac{a^2}{b^2} \right) \right]^{1/2} \tag{2}$$

where $a$ is the minor radius of the ellipse and $b$ is the major radius of the ellipse.

The natural frequencies of a moving pool are given by

$$f_n = 203 \frac{c_n}{w} Hz \tag{3}$$

where $c_n$ is dependent on eccentricity $\varepsilon$ and is plotted in [4].

## b. Model Two for Partial and Full Penetration Pool

The observed oscillation behaviour of both the partially penetrated and the fully penetrated weld pool can be explained in terms of classical hydrodynamics, taking into account the pressure balance on the weld pool surface.

### Partial Penetration

In the case of the partially penetrated weld pool, two different oscillation modes (modes 1 and 2) can occur. The observed oscillation behaviour can be described mathematically by applying the principles to the liquid metal in the weld pool, assuming the liquid metal in the weld pool to be incompressible. The frequency of the two oscillation modes can be calculated as [5]:

$$f = 0.07 D_1^{-3/2} \qquad f = 0.04 D_1^{-3/2} \qquad\qquad (4)$$

with $D_1$ the diameter of the circle which has a surface area equal to the surface area of the weld pool.

## Full Penetration

The situation of the full penetration pool is essentially different from that of the partial penetration pool. Whereas the liquid weld metal in the partial penetration pool is backed by a solid bottom, this is not the case in the full penetration pool. In fact, the liquid weld metal in the full penetration pool has an extra degree of freedom (normal to the surface of the pool) and it is evident that this will influence the oscillation behaviour. The fully penetrated weld pool can be regarded as a stretched membrane, the motion of which is controlled by the surface tension of the two surfaces. Oscillation occurs in mode 3. Its oscillation frequency can be described as [5]:

$$f = 0.2 D_2^{-1} \qquad\qquad (5)$$

with $D_2$ the diameter of the equivalent cylinder, the volume of which equals the volume of the fully penetrated weld pool.

### 2.1.3    Pool Oscillation Detection

Weld pool oscillation frequency can be measured by means of high-speed photography. It also can be measured by monitoring the arc radiation, making use of the phenomenon that the intensity of the arc radiation is proportional to the arc length. But most frequently, and especially for the Gas Tungsten Arc Welding (GTAW) process, since the pool is forced mainly by the arc plasma, the pool motion can be detected using the arc voltage. No additional instrumentation would be necessary to accomplish this sensing. Therefore, the promise of pool dynamics as a sensing method is to provide topside pool geometrical information simply by modulating the current and performing signal processing on the resulting arc voltage.

The natural frequencies of the weld pool can be inferred from the power spectrum estimation of the arc voltage-current relationship. Thus these models can act as estimators or observers that infer geometry information from the available measurements.

### 2.1.4    Results

The presence of the predicted peaks in the frequency spectrum has been validated by experiments. The typical voltage power spectrum diagram (PSD) plot [4] displays many peaks, several of which are near the frequencies predicted by model 1. The converse prediction about the weld pool back width based on pool oscillation frequency can also be concluded as:

$$f_0 = -2.1 + \frac{176}{w} \qquad f_1 = 3.1 + \frac{213}{w} \tag{6}$$

where $f_0$ and $f_1$ are the frequencies of the first and second mode of model 1 in Hz and $w$ is the measured back pool width in mm.

From model 2, it may also be expected that the translation from partial penetration to full penetration will give rise to an abrupt change in oscillation behaviour. It is believed that this transition can be used in process monitoring and control of weld penetration. It was found that during the growth of the weld pool, the oscillation frequency at first decreased gradually from 336 to 204 Hz in the range of partial penetration, then drops abruptly after reaching a certain degree of penetration and finally decreases again slowly from 100 to 37 Hz in the range of full penetration. It appears that the sharp drop in oscillation frequency does not exactly coincide with the transition from partial penetration to full penetration. The sharp transition of oscillation occurs at a value of weld pool back width greater than zero (that means the weld pool is still not fully penetrated) due to the high pressure action of the arc on the weld pool [3].

### 2.1.5    Disadvantages

Even though experiments have validated the presence of the predicted peaks in the frequency spectrum of arc voltage, these peaks are often obscured in the noisy process. The proximity of all the peaks makes a unique identification unreliable without other correlating data. The variance/resolution trade-off inherent in this measurement scheme makes it only good for a rough measurement. Also, the time required for calculating the geometry parameters may cause too great a delay in the control scheme if this measurement was the only one available. Finally, what may be considered the greatest limitation with this measurement scheme is the fact that the voltage component from the weld pool motion is actually an average drop across the whole pool. Further research is necessary in developing a measurement scheme which would reflect either the motions of the pool centroid, or better still, those at various points in the pool surface. The measurement of pool impedance through analysis of the arc voltage may have the above problems associated with it. A better approach may be to directly measure the motion of the pool surface using high-speed video means.

## 2.2     *Ultrasound*

### 2.2.1     *Contact Transducer*

Previous research concerning the investigation of weld quality by means of ultrasound has relied primarily on contact or immersion transducers for both the generation and reception of ultrasound [6-8]. For example, a 5 MHz immersion transducer, generating longitudinal and mode-converted shear waves, was used in [6]. The transducers operate in a pulse-echo mode. Initially, the transducer acts as a source of sound, sending an ultrasonic pulse through the weld pool. The sound reflects from an interface (especially solid-liquid interface) where the density or sound speed changes. The transducer then acts as a receiver and detects these echoes. The received echo amplitude is plotted versus the time of arrival (such a plot is called an A scan). The time between the initial pulse and the arrival of the echoes is related to the distance of the interfaces and the sound speeds. Thus, if the sound speeds in the various media between the transducer and the interfaces are known, the distance to the interfaces can be calculated.

Contact transducers have been used successfully for post-process inspection of welds, and have proven highly reliable for this purpose. Also, contact transducers are capable of generating strong signals and, compared with non-contact devices, such as electromagnetic acoustic transducers (EMAT) or interferometers, are less noisy as receivers.

Researchers attempting to use contact transducers for in-process monitoring of welds have found the hot surface of the test specimen to be a problem for continuous operation of the contact probes. Radiation from the welding arc can be damaging for the long-term operation of the transducers, even if the devices are shielded. Additionally, contact devices, custom-made to allow for motion along the specimen surface, are limited to flat surfaces where even slight surface imperfections can cause "kick-up" of the transducer, resulting in a momentary loss of signal. Clearly, investigation of a non-contact means of generating and receiving ultrasound is required for industrial welding applications.

### 2.2.2    Laser Array and EMAT Ultrasonic Measurement

#### a. Experimental Set-up

Standard ultrasonic techniques require the transducer to be mechanically coupled to the test specimen, and are therefore unsuitable for real-time applications where samples are in motion, at elevated temperatures, or in harsh environments. Lasers and electromagnetic acoustic transducers provide non-contact means for generating and receiving ultrasound and should therefore be considered where traditional contact transducers cannot be employed.

A laser probe and an EMAT are configured on opposite sides of the weld, rather than generating and receiving ultrasound in the centre of the weld itself. Figure 1 shows the experimental set-up used for the measurement of penetration depth in simulated liquid welds with laser generation and an EMAT receiver.

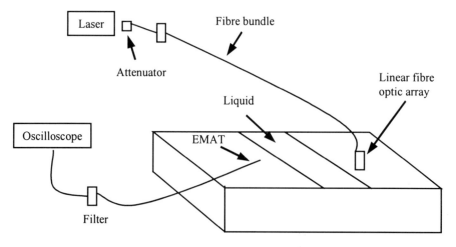

Figure 1 Set-up for penetration depth measurement in simulated liquid welds.

Simulated liquid welds, rather than actual welds, were studied for a number of reasons. Actual welds have been shown to yield inconsistent results despite efforts to keep all parameters constant. Although simulated welds do not accurately represent the complexity of the real weld geometry and the physical process, they allow the experiment to be more fully controlled and will provide baseline data for comparison with subsequent real weld studies. Also, simulated welds can be more speedily prepared and with less expense, allowing a variety of sensor configurations to be easily investigated. The primary purpose of this experiment is to assess the feasibility of applying these inspection techniques for real weld control.

## b. Theoretical Predictions of Penetration Depth

Figure 2 shows the configuration for sensing penetration depth in simulated weld pools.

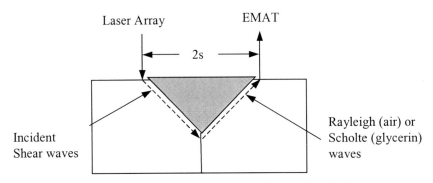

Figure 2 Configuration for sensing weld penetration depth.

The shear wave generated by the laser array travels through the steel plate and is incident on the bottom of the simulated weld pool where it is mode-converted into a Rayleigh (air filled vee-groove) or Scholte wave (glycerin filled vee-groove). When the groove is empty (steel-air interface), the Rayleigh wave continues around the perimeter of the groove until it reaches the surface of the plate on the opposite side of the groove where it is detected by the EMAT. When the groove is filled with glycerin, the incident shear wave is converted into a Scholte wave which travels up the perimeter of the pool until it reaches the surface of the plate where it continues as a Rayleigh wave.

Time-of-flight methods are used to calculate the penetration depth. Time of flight is related to the wave speeds for a 60 degrees air filled vee-groove by Equation (7) [10]. Equation (8) [10] gives the time-of-flight for a 60 degrees vee-groove filled with glycerin where each wave speed is a function of temperature. These equations could be numerically solved for penetration depth (PD).

$$t = \frac{\sqrt{s^2 + PD^2}}{c_T(T)} + \frac{\left(s - PD\tan 60° + \frac{PD}{\cos 60°}\right)}{c_R(T)} \tag{7}$$

$$t = \frac{\sqrt{s^2 + PD^2}}{c_T(T)} + \frac{\left(\frac{PD}{\cos 60°}\right)}{c_H(T)} + \frac{\left(s - PD\tan 60°\right)}{c_R(T)} \tag{8}$$

where $2s$ is the distance between the laser array and the EMAT, $PD$ the penetration depth of the simulated weld pool, $t$ the time of flight, $c_T$ the shear wave speed, $c_R$ the Rayleigh wave speed, and $c_H$ the Scholte wave speed.

Time of flight can be obtained from the experiments. Thus, these equations were used to plot the theoretical curves for penetration depth.

## c. Results

The results for glycerin filled grooves showed that the time of flight increased with increasing penetration depth. The trend in the experimental data agrees with theoretical predictions. This method was shown to have an error of less than 7 percent for all penetration depths and temperature profiles tested. It showed that this was a reliable method for the measurement of penetration depth in simulated liquid weld pools by ultrasound time of flight techniques. In real welds the glycerin would be replaced by liquid steel, in this case it will be necessary to adjust the Scholte wave speed to that for a solid and liquid steel interface and temperature gradients should be accounted for when assessing theoretical time of flight values.

## d. Disadvantages

The main disadvantage is that all the tests are done in simulated welds. Several modifications to the apparatus described in this chapter need to be considered before attempting to measure penetration depth of real welds. First, both attenuation and noise would be higher in real weld experiments. Because the specimen is at an elevated temperature in the manufacturing setting, a stronger laser source (a phased array system) is recommended, which can increase the magnitude of ultrasound generation and can focus the energy in the desired direction. Additionally, the optical fibres should be replaced with high temperature fibres and protected from the shielding gas and other debris associated with welding.

## 2.3 *Infrared Sensing*

### 2.3.1 *Theoretical Foundations*

The weld pool can be treated as a blackbody. Thus, its maximum emission wavelength can be described as:

$$\lambda_{max} T = 0.2898 \times 10^{-2} \, mK \tag{9}$$

Weld pool temperature varies from melting point to boiling point of the welded materials. Suppose it is from 1773 K to 2273 K for mild steels, thus it could be concluded that $\lambda_{max}$ is from 1.27 µm to 1.63 µm, which is in the infrared range.

The intensity of the weld pool emissions as a blackbody is represented as follows:

$$I(\lambda, T) = 2\pi h c \left/ \left[ \lambda^5 \left( e^{hc/\lambda kT} - 1 \right) \right] \right. \tag{10}$$

where $h$ is Plank constant, $k$ Boltzman constant, and $c$ the speed of light.

Suppose $T$ is equal to 2000 K, the relationship between the relative intensity and spectrum of weld pool emissions is shown as in Figure 3. It shows that weld pool emissions are mainly in the infrared ranges and infrared sensing can provide another approach of sensing the weld pool geometry.

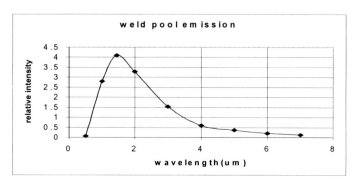

Figure 3 Theoretical weld pool emissions.

## 2.3.2    *Infrared Sensing Technology*

With the advancement of the infrared (IR) technology, IR detectors have developed from liquid nitrogen cooled to non-cooled, from mechanical scanning to Focal Plane Array systems, from low speed to high speed "live" ones. All of these triggered the release of high performance IR cameras that offer significantly improved sensitivity, image quality, detector reliability and noise reduction. Table 1 lists the main specifications for an infrared camera.

Table 1 Specifications of an IR camera.

| Detector | PtSi/CMOS 256x256 FPA with variable integration |
|---|---|
| Image update rate | 60/50 $Hz$ |
| Spectral band | 3.4 to 5 μm |
| Sensitivity | 0.07 °C |
| Accuracy | ±2 °C |
| Temperature measurement range | -10 to 450 °C extendable up to 2000 °C |

The infrared image processing software has advanced greatly. Information about the images such as isotherms, weld pool areas, and cross section temperature profile, is readily available.

## 2.3.3    *Results*

The test was carried out in Gintic Institute of Manufacturing Technology. An IR camera, Varioscan, was used to capture the weld pool temperature distribution profile. The working temperature range were defined from 200 °C to 1500 °C, thus the interference from both the environment and the arc was eliminated. Spot welding was performed by a Fronius TT 2000 TIG welding power source. Figure 4 shows the infrared images during spot welding at different times. The cross section temperature profile is illustrated in Figure 5.

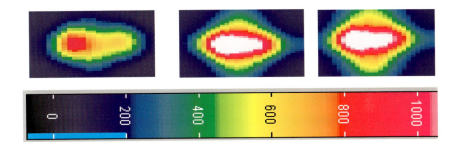

Figure 4 Weld pool temperature distribution at different times during spot welding.

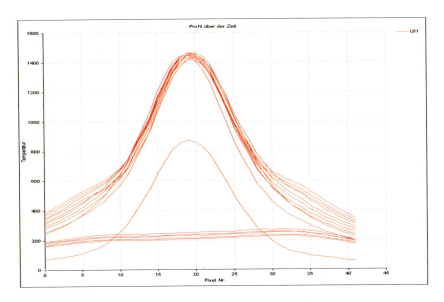

Figure 5 Cross section temperature profiles.

## 2.3.4    Limitation

The principal limitation to this type of sensing is its currently prohibitive cost. However, new inexpensive CCD sensors (charge coupled devices) are currently being developed, and should be commercially available for sensing applications in the foreseeable future.

## 2.4    *Surface Depression Sensing*

### 2.4.1    Background

Strictly speaking, no flat weld pool surfaces exist in arc welding processes. Pool surface deformation can always be observed in gas metal arc welding (GMAW) and gas tungsten arc welding (GTAW) with filler, due to the mass transfer. For GTAW without filler, pool surface deformation is also apparent in the full penetration mode. In the case of partial penetration, three modes of pool deformation can be observed for different current levels. Thus, the deformation of the pool surface is an inherent characteristic of arc welding processes, and an important phenomenon because of its influence on the arc energy distribution and its correlations with possible weld defects and weld penetration.

### 2.4.2    Sensing Technology

Pioneering work was conducted at the Ohio State University, using radiography [16]. The radiation of the received x-ray increases with the depression depth. Using this approach, many valuable results have been obtained. However, only the case of stationary arc was addressed. To avoid the interference of electrode and gas nozzle, long electrode extension and inclined torch attitude were used. The imaging device and X-ray source could not both be attached to the torch to form a so-called topside sensor. This fact, in addition to the radioactivity, restricts the prospective application of this method in practical monitoring or control of weld pools. Also, the principle behind this method is to measure the material thickness. For the case of full penetration where the backside pool surface deformation occurs, the pool surface shape will be difficult to extract.

Due to the difficulty involved in on-line pool surface deformation sensing, the geometry of the weld behind the pool was measured as an alternative [17-18]. It was found that the average weld depression depth, which is defined as the cross weld depression area divided by the width, has a close relationship with the full penetration state which is specified by the backside bead width. A corresponding closed-loop control system has been developed, and satisfactory control has been obtained. However, there is an inherent measurement delay if the weld behind the pool is monitored rather than the pool itself. If the pool surface deformation can be monitored, a promising weld penetration control strategy can be expected.

A novel mechanism for observing the pool surface shape was proposed. Laser stripes projected through a novel grid are reflected from and deformed by the mirror like pool surface [19-20]. To acquire three-dimensional weld pool surface information, a special technique must be implemented. The common practice for determining the three dimensional shape of a surface is to project a structured light on to the surface and sense the diffuse reflection of the structured light. The shape information can then be extracted from the deformation of the structured light. However, the weld pool surface is mirror like smooth and no substantial amount of structured light can be reflected diffusely. It appears that only specular reflection can be utilised.

The proposed incident mechanism of structured light is realised through a specialised grid. This grid consists of a common grid and frosted glass. When the laser is projected on to the frosted glass, the laser travel direction will be changed. From the viewpoint of light travel, any point on the frosted glass can be considered as a new point light source which disperses light with a certain diffuse angle. The camera views the slits (grid openings) through their reflection from the weld pool surface. Their virtual image consists of bright stripes deformed by the weld pool surface deformation and is sensed by the camera. The surface shape of the weld pool is contained in the acquired image.

## 2.4.3    Image Processing

Although a human can identify the boundary of the weld pool from the acquired images, machine recognition may not be straightforward because of the complexity of the acquired image. In the acquired image, different scenes exist: torch, electrode, oxidised heat affected zone, edges, laser stripes and boundary of the weld pool. To distinguish between these scenes; their inherent features, optical, geometrical or locational, must be found. Using locational and optical features and through edge detection, boundary fitting and segmenting, boundaries of torch, electrode, laser stripes, heat-affected zone (HAZ) and weld pool have been successfully identified.

## 2.4.4    Results

In general, the results of the image processing are quite satisfactory. The parameters of the fitted models for the stripe edges and pool boundary provide a description of the real pool surface. Using these parameters, the free form weld pool surface can be reconstructed based on the reflection law.

Because of its unique correlation with the weld pool surface, the reflection pattern can be directly used to control the weld penetration without surface shape calculation.

## 2.4.5    Disadvantages

Currently, a period of one second is needed to obtain the weld pool surface from the acquired image. In order to apply this technology to real-time welding control, the imaging and processing speeds have to be significantly increased.

It could be concluded that pool oscillation sensing, ultrasound sensing, infrared sensing and depression sensing all have their own constraints. The most promising sensing approach is the direct topside weld pool observation.

In the remaining of this chapter, a vision-based weld pool geometry control system using neurofuzzy logic will be discussed in detail. In this system, both the weld pool length and width were used to realise welding fusion control. The weld pool images were captured by a topside pulse laser assisted high-shutter-speed camera – LaserStrobe [26]. Two real-time image processing algorithms were developed to detect the weld pool boundary from the captured images in less than 50 ms. A neurofuzzy controller was developed for the real-time control of the GTAW welding process. In such a controller, experimental results about the relationships between welding parameters and weld pool dimensions were used to adjust rules of the fuzzy model by learning techniques developed in the neural networks. Such a control system will combine the advantages of fuzzy system, such as, the transparent representation of knowledge and the ability to cope with uncertainties, as well as the advantages of neural networks, such as, the ability to learn.

# 3.   Weld Pool Vision and Control System

## 3.1   *Controlled Welding Process*

In arc welding there are a lot of interactions among heat, arc and arc force. The welding process is physically complex. The heat input of the arc in a unit interval along the travel direction can be written as:

$$\Delta H \propto (i \ / v)u \tag{11}$$

where *i* is the welding current, *v* is the welding speed, and *u* is the welding voltage. Roughly speaking, one can assume that the area of the weld pool is approximately proportional to *ΔH*. A lot of other factors, such as surface conditions, also play a significant role in achieving good welding quality. Hence, the correlation between the weld pool geometrical dimensions and the welding variables is non-linear. Furthermore, because of the thermal inertia, the process is highly dynamic.

The major adjustable welding parameters include the welding current, arc length, and travel speed of the torch. In general, the weld pool increases as the current increases and the travel speed decreases. When arc length increases, the arc voltage increases so that the arc power increases, but the distribution of the arc decreases. As a result, the correlation between the weld pool and arc length may not be straightforward. In addition to these three welding parameters, the weld pool is also determined by the welding conditions such as the heat transfer condition, material, thickness, and chemical composition of the workpiece, shielding gas, angle of the electrode tip, etc. In a particular welding process control system, only a few selected welding parameters are adjustable through the feedback algorithm to compensate for the variations in the welding conditions. Compared with the arc length, the roles of the welding current and welding speed are much more significant and definite. In the present study, the welding speed can be adjusted on line through the robot. Thus, in addition to the welding current, the welding speed was selected as another control variable. The controlled process can therefore be defined as a GTAW welding process in which the welding current and speed are adjusted on line to achieve the desired topside width and length of the weld pool.

## 3.2   *Weld Pool Vision System*

The difficulties of extraction and observation of weld pool are due to the interference from the welding arc. To eliminate this interference, a strong illumination must be projected onto the area of interest. The ultra-high shutter speed vision system (LaserStrobe) is ideal for this application [26].

LaserStrobe consists of a pulsed laser illumination unit, a camera head and a system controller. Firstly there is a band pass optical filter in the camera (300 nm ~ 350 nm, the laser wavelength is 337.1 nm) to cut off arc emissions in other wavelengths.

Secondly the pulse duration of the laser is 5 ns, with which the camera is synchronised and the laser power density in the duration is very high. Thus the intensity of laser illumination is much higher than that of the arc during the pulse.

Thirdly the reflection from the molten weld pool to the laser is much lower than that from the solidified weld bead. Thus the contrast between the weld pool and its surrounding areas is also improved. Using this vision system, good weld pool images can be obtained under different welding conditions.

## 3.3   *Set-Up of Weld Pool Vision and Control System*

The experimental set-up consisted of a 4-axis AC servo motor robot, a welding torch, and the LaserStrobe camera all linked to a Pentium I computer. The computer contained a DT 3152 frame grabber board, an 8-axis motion controller card called the NextMove PC controller and the fuzzyTECH [27] and VisionBlox software [28], which made up the data acquisition and robotic control system.

Figure 6 is a block diagram showing the various components of the system and their relationships. The CPU sends signals to the NextMove card, instructing it to move the slider while welding is carried out on top of the slider. The camera attached to the welding torch captures images of the weld pool formed, and sends them to the DT 3152 frame grabber board. From there the images are sent to the VisionBlox software that detects the width and length of the weld pool.

The width and length of the weld pool are passed on to the fuzzy controller, which outputs control measures to be carried out respectively by the motion controller card and the welding power source interfaced with the computer. The NextMove card thus sends a signal to the slider to speed up or slow down its movement. The welding power source interface with the computer sends a signal to increase or decrease welding current. Figure 7 shows the front view of the physical system.

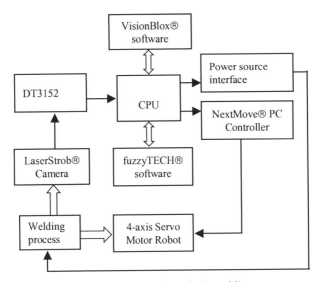

Figure 6 Block diagram of the robotic welding system.

Figure 7 Front view of hardware components.

# 4.      Weld Pool Geometry Extraction

## 4.1    *Real-Time Weld Pool Geometry Extraction*

Some typical images of GTAW weld pool, captured by the above mentioned vision system, are shown in Figure 8.

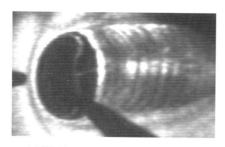 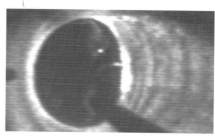

(a) Weld pool at I=100A, V=6 mm/s.            (b) Weld pool at I=130 A, V=6 mm/s.

Figure 8 Weld pool images at different welding parameters.

To propose an effective algorithm, the image features must be analysed. It can be seen that the greyness difference between the weld pool and its surrounding solid material appears as the most apparent feature of the image. The greyness values associated with the molten weld pool are low due to its high absorption of the laser. The greyness of the surrounding solidified areas was much higher because of their high reflection to the laser.

## 4.2    *Edge Detection*

One approach was to use the Caliper Tool of VisionBlox [28], which was used to find one or more edge pairs within a region of interest (ROI) set on the binary weld pool images. An edge is a point on the image where the contrast between the two sides is sufficiently high. The polarity of these changes in intensity, either from dark to light or light to dark can also be specified in order to narrow down the search. When the edges are found, the Caliper Tool can be used to find the separation between them.

The region of interest for the width and length of the weld pool was set such that it selected the region where there is no overlapping of the shadow of the electrode and the weld pool itself. Similarly, the region of interest was chosen such that it was placed where high reflection was less likely to occur. High reflection was due to welding flux floating on weld pool. The resulting image and weld pool dimensions are displayed in Figure 9.

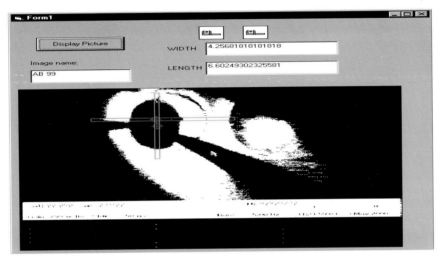

Figure 9 Weld pool image and dimensions after edge detection.

However, the proposed method was not as effective in avoiding reflective surfaces. The edge of the reflective surface can sometimes be taken for the edge of the weld pool, making the width or length value obtained to be less than the actual measurement.

### 4.3    *Connectivity Analysis*

An alternative approach was to use the Blobs Tool of VisionBlox to get the dimensions of the weld pool [28]. Blobs are areas with similar intensity. The software performs a connectivity analysis on the inspection image and displays the results. The purpose of the connectivity analysis is to segment the inspection image into areas of similar intensity. Information concerning the blobs, such as length and breadth, is then available.

Before performing blob analysis an area of interest is set around the weld pool to shorten image-processing time. In our study, the image processing time is about 50 ms. The processed image and the results are shown in Figure 10 and Figure 11.

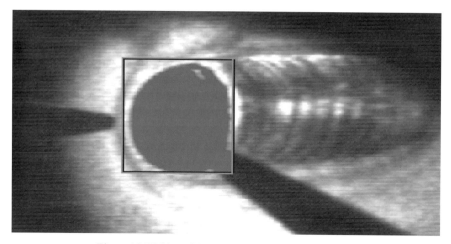

Figure 10 Weld pool image after connectivity analysis.

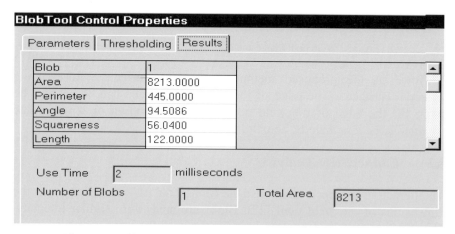

Figure 11 Weld pool geometry information after connectivity analysis.

## 4.4 *Relationship between Weld Pool Dimensions and Welding Parameters*

Since weld pool geometry was available, bead-on-plate welding was conducted to explore the relationship between the weld pool geometry and welding parameters. Some of the results are shown in Figure 12 and Figure 13. These data were used to train the rule base of neurofuzzy logic control system.

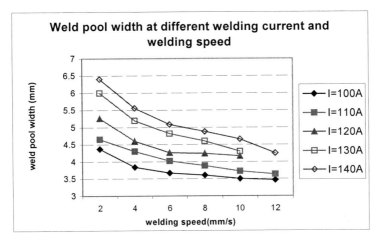

Figure 12 Correlation between weld pool width and welding current and speed.

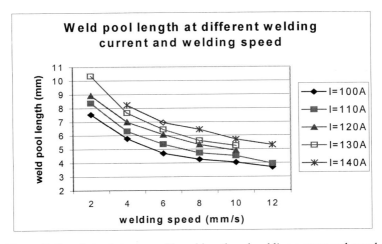

Figure 13 Correlation between weld pool length and welding current and speed.

## 5.    Neurofuzzy Logic Control

### 5.1    *Fuzzy Logic Control*

It is commonly accepted that it is difficult to find suitable analytical models to describe a welding process. Details of the process is influenced by operator's experiences. Fuzzy logic control, however, offers certain advantages, such as: the transparent representation of human knowledge and the ability to cope with uncertainties. Therefore in this study, a fuzzy logic control system was used to control welding process on line by emulating human's decision-making process.

A fuzzy system has three major conceptual components: rule base, database, and reasoning mechanism. The database contains the membership functions of the fuzzy sets. In this study, we defined weld pool length and weld pool width as the two input linguistic variables. The range of weld pool length was from 4 mm ~ 18 mm. Its fuzzy sets were *Small, Medium, High*, and *Very High*. The range of weld pool width was from 3.5 mm ~ 6.5 mm. Its fuzzy sets were *Small, Medium, High*, and *Very High*. We also defined the change of the welding current and the change of the welding speed as the two output linguistic variables. The range of the change of the welding current was –20 A ~ +20 A. Its fuzzy sets were also *Small, Medium, High*, and *Very High*. The range of the change of the welding speed was –4 mm/s ~ +4 mm/s. Its fuzzy sets were *Small, Medium, High*, and *Very High*, too. The membership functions could be defined as Z-type, Pi-type, Lambda-type, S-Type or corresponding to an arbitrary shape.

The rule base consists of the conventional fuzzy IF-THEN rules. These rules are elicited from human knowledge such as:

- If the length of the weld pool is low and the weld pool width is low, then the change of the current will be high and the change of the speed will be low.
- If both the length and width of the weld pool are high, then the change of the current is medium and the change of the welding speed will be high.

The reasoning process includes fuzzification of crisp input values, fuzzy rule inference and defuzzification to calculate the crisp output value.

## 5.2    *Neurofuzzy Logic System*

In conventional fuzzy models, the fuzzy linguistic IF-THEN rules are primarily derived from human experience. That means that operator's experience is the major source to establish the fuzzy model. In these models, no systematic adjustments are made on the used rules, membership functions, or reasoning mechanism based on the behaviour of the fuzzy model. In general, if the fuzzy rules elicited from the operator's experience are correct, relevant, and complete, the resultant fuzzy model can function well. However, frequently such fuzzy rules from the operators do not satisfy the correctness, relevance, and completeness requirements; the rules may be vague and misinterpreted, or the rule base could be incomplete. In such cases, the performance of the fuzzy system can be greatly improved if systematic adjustments are made based on its behaviour.

Information about welding processes is also contained in experimental results. When knowledge about the system to be designed is contained in data sets, a neural net presents a solution since it can train itself from the data. So neural net can learn from data sets, while fuzzy logic solutions are easy to verify and optimise. Neurofuzzy is a combination of fuzzy logic control and neural networks [27]. The term neurofuzzy modelling is used to refer to the application of algorithms developed through neural network training to identify parameters for a fuzzy model. Thus a neurofuzzy model can be defined as a fuzzy model with parameters, which can be systematically adjusted by training algorithms implemented in neural networks based on experimental results.

In each of the three computation mechanisms including fuzzification, fuzzy inference and defuzzification, we must find parameters that can be used as "weights" in a neural training procedure. Second, the error backpropagation algorithm differentiates the transfer functions of the neurons to determine the influence of each neuron. Here, the standard fuzzy logic inference cannot be differentiated.

To solve the first problem, a fuzzy logic rule with an associated weight, or degree of support (DOS) is used [27]. Implementing a modified error backpropagation algorithm with fuzzy logic [27] can circumvent the second problem.

First, a conventional fuzzy logic control system, as shown in Figure 14, was established. The two input parameters were weld pool width and weld pool length. The two output parameters were the change of welding current and the change of welding speed. The rule base was developed from operator's knowledge.

Secondly, the experimental data were formatted and saved in a file and then used to train the degree of support for each rule. Figure 15 shows the training process.

The rule base after training is shown in Figure 16. It can be seen that each rule is given a degree of support.

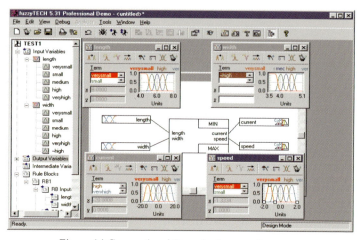

Figure 14 Conventional fuzzy logic control system.

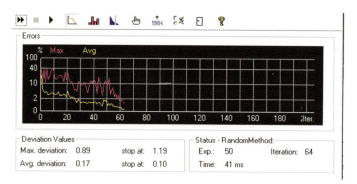

Figure 15 Training process of rule base.

| # | IF | | THEN | | THEN | |
|---|---|---|---|---|---|---|
| | length | width | DoS | current | DoS | speed |
| 1 | verysmall | verysmall | 1.00 | veryhigh | 0.60 | veryhigh |
| 2 | verysmall | small | 0.80 | veryhigh | 0.80 | small |
| 3 | small | verysmall | 0.80 | veryhigh | 0.80 | small |
| 4 | small | small | 1.00 | high | 0.60 | small |
| 5 | small | medium | 0.80 | high | 0.50 | high |
| 6 | medium | small | 0.80 | high | 0.50 | high |
| 7 | medium | medium | 1.00 | medium | 1.00 | medium |
| 8 | medium | high | 1.00 | small | 0.60 | small |
| 9 | high | medium | 1.00 | small | 0.60 | small |
| 10 | high | high | 0.80 | small | 0.80 | high |
| 11 | high | veryhigh | 0.70 | verysmall | 0.80 | veryhigh |
| 12 | veryhigh | high | 0.70 | verysmall | 0.80 | veryhigh |
| 13 | veryhigh | veryhigh | 0.80 | verysmall | 0.80 | veryhigh |
| 14 | | | | | | |
| 15 | | | | | | |
| 16 | | | | | | |
| 17 | | | | | | |
| 18 | | | | | | |

Figure 16 Rule base after training.

## 5.3 System Optimisation and Integration

There are some other methods to optimise the neurofuzzy control system, such as, 3D plot (Figure 17), time plot and rule analyser. The 3D plot helps analysis of the static input and output characteristic of a neurofuzzy system as a scaleable, rotational 3D plot. The time plot assists analysis of the time response characteristics of a fuzzy logic system. The rule analyser shows which rules are fired.

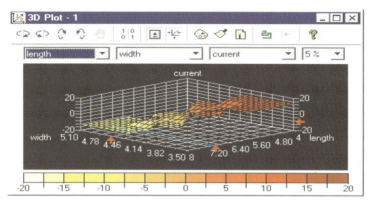

Figure 17 3D plot for optimisation.

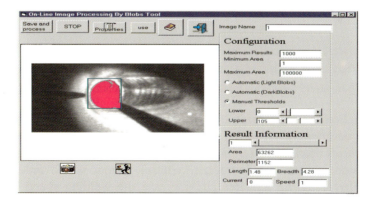

Figure 18 Interface of the whole system from image process to neurofuzzy control.

Finally image capturing, image processing and neurofuzzy logic control were integrated into one system. The output of image processing is the input to the neurofuzzy logic control system, and then the neurofuzzy logic control system gives the values of the changes in welding current and welding speed to the welding power source and robot. Figure 18 shows the interface of the integrated system.

## 6.    Results

### 6.1    *Simulation Results of Neurofuzzy Logic Control System*

The calculation of the neurofuzzy control system is the same as a conventional fuzzy logic control system, involving three steps:

- Fuzzification of crisp input values.
- Fuzzy rule inference. It includes aggregation of the IF part of a fuzzy rule, composition of the THEN part of a fuzzy rule, and result aggregation.
- Defuzzification to calculate the crisp output value.

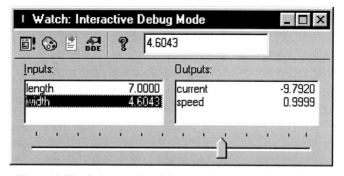

Figure 19 Simulation results of the neurofuzzy control system.

To simulate the neurofuzzy control system, the weld pool length and weld pool width were set to 6.0 mm and 4.5 mm respectively. When the weld pool sizes changed to 7.0 mm in length and 4.6 mm in width, detected by the vision system, the neurofuzzy control system would reduce the welding current by 9.8 amperes and increase the welding speed by 1.0 mm/s (Figure 19).

### 6.2    *Closed-Loop Control of Welding Speed*

To test closed-loop control on the width and length of the weld pool using fuzzy control, a two-input (weld pool width and length) and one-output (change of welding speed) neurofuzzy controller was designed. The change in speed causes the weld pool geometry to change as well. The closed-loop control system is represented in Figure 20. From this diagram, the vision system was seen as the feedback mechanism, the actual width and length were compared to the desired values and the error was input into the fuzzy controller.

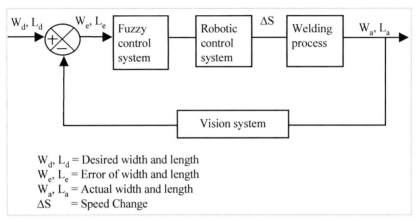

Figure 20 Block diagram of the closed-loop control system.

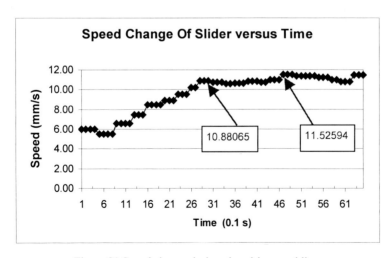

Figure 21 Speed change during closed-loop welding.

First, open-loop welding tests were conducted at a welding current of 115 A and a welding speed of 6 mm/s to identify the average width and length of the weld pool. The average width was found to be 5.037 mm, and the average length 5.919 mm. Then the desired width and length were set to 10% below the average values (4.5 mm and 5.3 mm). The objective of this experiment was to observe if the fuzzy controller could speed up the robot in order to produce the desired weld pool geometry. The resulting speed changes are shown in Figure 21.

It was verified that the speed steadily increased until it reached a stable state (10.88 mm/s ~ 11.0 mm/s). It was also found that the average width and length were 4.655 mm and 5.322 mm respectively. These values were found to be close enough to the desired values, demonstrating that fuzzy control of weld pool operated effectively. Open-loop and closed-loop weld beads are shown in Figure 22.

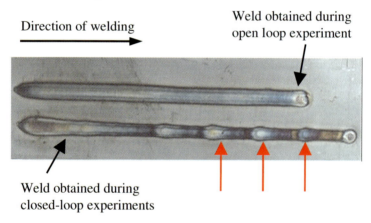

Figure 22 Weld beads during open-loop and closed-loop welding experiments.

In Figure 22, the three arrows at the right bottom point to places in the actual weld where larger weld pools were formed. These larger-than-usual weld pools are not caused by incorrect responses of the fuzzy logic controller. Rather, it is believed that the processor, a Pentium-I 200 MHz chip, in the computer, which houses all the software and the hardware cards has slowed down at these points while doing the complex calculations required by both the image processing and fuzzy control modules. The delayed responses from the CPU might cause the robotic slider to pause while waiting for instructions from the CPU. Since the torch is always activated during the entire welding session, the slider's pause will allow more metal to be heated, causing the larger-than-usual weld pools seen in Figure 22.

It is recommended that a faster computer be used for doing the closed-loop control of the system. Ideally, the processor should have at least a Pentium III processor, with enough random access memory (RAM) to carry out the computation-intensive image processing functions

## 7. Conclusions

The study has demonstrated the advantages and feasibility of weld pool sensing with the vision system and weld pool geometry control using the neurofuzzy control technique. The following conclusions can be drawn from the study:

1. Relatively clear weld pool images can be captured using the laser illuminated vision system to eliminate the interference from the welding arc.
2. Image processing based on connectivity analysis provides the measurement of the weld pool geometry on line.
3. A neurofuzzy logic control system combining both the advantages of a fuzzy system and those of a neural network is suitable for a complex process where an accurate mathematical model is not available.

Results from this study show that weld pool observation can be achieved with the help of the special high shutter speed camera such as LaserStrobe. The weld quality can be stabilised by neurofuzzy logic control. These techniques have great potential for many industrial applications where high welding quality is required.

## References

1. Zacksenhouse M., and Hardt D.E. "Weld pool impedance identification for size measurement and control", ASME Journal of Dynamic Systems, Measurement, and Control, 105(3), 1983, pp. 179-184.
2. Renwick R.J., and Richardson R.W. "Experimental investigation of GTA weld pool oscillations" Welding Journal, 62(2), 1983, pp. 29s-35s.
3. Xiao Y.H., and Ouden G. Den, "A study of GTA weld pool oscillation", Welding Journal, 69(8), 1990, pp. 289s-293s.

4.  Tam, A.S., and Hardt, D.E. "Weld pool impedance for pool geometry measurement: stationary and nonstationary pools", Journal of Dynamic Systems, Measurement, and Control, 111(12), 1989, pp. 545-553.

5.  Xiao, Y.H., and Ouden, G.D. "Weld pool oscillation during GTA welding of mild steel", Welding Journal, 72(8), 1993, pp. 428s-434s.

6.  Hardt, D.E., and Katz, J.M. Ultrasonic measurement of weld penetration. Welding Journal, 63(9), 1984, pp. 273s-281s.

7.  Carlson, N.M., and Johnson, J.A. "Ultrasonic sensing of weld pool penetration", Welding Journal, 67(11): 1988, pp. 239s-246s.

8.  Carlson, N.M., et al. "Ultrasonic NDT methods for weld sensing", Material Evaluation, 50(11), 1992, pp. 1338-1343.

9.  Yang, J., et al. "Ultrasonic weld penetration depth sensing with a laser phased array", Proceedings of 1994 ASME International Mechanical Engineering Congress, PED - Vol. 68-1, Manufacturing Science and Engineering, Chicago. IL, Nov. 6-11, 1994, pp. 245-254.

10. Graham, G.M., Charle Ume, I., and Hopko, S.N. "Laser array /EMAT ultrasonic measurement of the penetration depth in a liquid weld pool", Transactions of the ASME, Journal of Manufacturing Science and Engineering, 122(2), 2000, pp. 70-75.

11. Chen, W., and Chin, B.A. "Monitoring joint penetration using infrared sensing techniques", Welding Journal, 69(4), 1990, pp. 181s-185s.

12. Banerjee, P., et al. "Infrared sensing for on-line weld geometry monitoring and control", ASME Journal of Engineering for Industry, Vol. 117(8), 1995, pp. 323-330.

13. Beardsley, H.E., Zhang, Y.M., and Kovacevic, R. "Infrared sensing of full penetration state in gas tungsten arc welding", International Journal of Machine Tools and Manufacture, 34(8), 1994, pp. 1079-1090.

14. Nagarajan, S., Banerjee, P., Chen, W.H., and Chin, B.A. "Control of the welding process using infrared sensors", IEEE Transactions on Robotics and Automation, 8(1), 1992, pp. 86-93.

15. Lin M.L., and Eagar T.W. "Influence of surface depression and convection on arc weld pool geometry', Transport Phenomena in Materials Processing, ASME, New York, 1983, pp. 63-69.

16. Rokhlin S.I., and Guu A.C. "A study of arc force, pool depression, and weld penetration during gas tungsten arc welding", Welding Journal, Vol. 72(8), 1993, pp. 381s-390s.

17. Zhang, Y.M., and Kovacevic, R. "Dynamic estimation of full penetration using geometry of adjacent weld pools", ASME Journal of Manufacturing Science and Engineering, Vol. 119(11), 1997, pp. 631-643.

18. Zhang, Y.M., and Kovacevic, R. "Real-time sensing of sag geometry during GTA welding", Transactions of the ASME Journal of Manufacturing Science and Engineering, Vol. 119(2), 1997, pp. 151-160.

19. Kovacevic, R., and Y.M. Zhang, Y.M. "Sensing free surface of arc weld pool using specular reflection: principle and analysis", Journal of Engineering Manufacture, Vol. 210, 1996, pp. 553-564.

20. Zhang, Y.M., and Kovacevic, R. "Real time image processing for monitoring of free weld pool surface", Transactions of the ASME Journal of Manufacturing Science and Engineering, 119(2), 1997, pp. 161-169.

21. Zhang, Y.M., Li. L., and Kovacevic, R. "Adaptive control of full penetration GTA welding", IEEE Transactions on Control Systems Technology, 4(4), 1996 pp. 394-403.

22. Henderson, D.E., Kokotovic, P.V., Schiano, J.L., and Rhode, D.S. "Adaptive control of temperature in arc welding", IEEE Control Systems Magazine, 13(1), 1994, pp. 49-53.

23. Zhang, Y.M., and Li. L. "Interval model based robust control of weld joint penetration", Transactions of the ASME Journal of Manufacturing Science and Engineering, 121(3), 1999, pp. 425-433.

24. Zhang, Y.M., Li. L., and Kovacevic, R. "Neurofuzzy model based control of weld fusion zone geometry", Proceedings of the American Control Conference. Vol. 4, IEEE, Piscataway, NJ, USA, 1997, pp. 2483-2487.

25. Zhang, Y.M., Li. L., and Kovacevic, R. "Neurofuzzy model based non-linear modelling and control of weld fusion zone geometry", Advanced Materials: Development, Characterization Processing, and Mechanical Behaviour. American Society of Mechanical Engineers, Materials Division (Publication) MD. Vol. 74, ASME, New York, NY, USA, 1996, pp. 149-150.

26. Hoffman, T. "Real time imaging for process control", Advanced Material and Processes, 140(3), 1991, pp. 37-43.

27. "FuzzyTech 5.3 User's Manual", Inform Software Corporation 1999.

28. "Vision Blox2.2 User's Manual", Integral Vision.

29. Luo, H., Devanathan, R., Wang, J., and Chen, X.Q. "Vision based weld pool geometry control using neurofuzzy logic", Proceedings of the 4th Asian Conference on Robotics and its Applications (ARCA), 6-8 June 2001, pp. 141-146.

30. Chen, X.Q., and Huang, S.S. "An investigation into vision system for automatic seam tracking of narrow gap welding process", Chinese Journal of Mechanical Engineering, Vol. 29, No. 3, June 1993, pp. 8-12.

# CHAPTER 7

# AUTOMATIC GTAW SYSTEM CONTROL AND TELEOPERATION

Andrew P. Shacklock, Wen Jong Lin, Hong Luo, and Sheng Huang

*Gintic Institute of Manufacturing Technology,*
*71 Nanyang Drive, Singapore 638075*

## 1.    Introduction – The Automatic Welding of High Performance Alloys

Gas Tungsten Arc Welding (GTAW) of titanium is a skilled task that is difficult to automate on account of the need to provide inert gas shielding for heat affected zones. Shielding fixtures and shielding chambers restrict or preclude human access to the welding process. Increased levels of automation are desired to overcome skilled labour shortages and to improve quality consistency. An automation system must possess some of the attributes of the skilled welder if it is to cope with the many variables, uncertainties and spurious signals encountered in this complex multi-parameter process.

Much effort has been put into research and development of automatic welding technology [1, 2, 3]. The work described here addresses the problems of integrating these technologies in a working system for GTAW of titanium. Firstly, the special process requirements of titanium are discussed as these constrain the system design. The features of a skilled automation system are presented in light of this. A major section is devoted to design and analysis of the welding manipulator with particular attention to its spatial kinematics. This is because welding must be performed in a confined work volume with problems of access and collision avoidance. A second reason is that precise and usable definitions of tool position and orientation are needed for sensorigeometric model fusion in a remotely operated system.

In the manual process, the welder controls the process using his sensory faculties, judgement, knowledge and experience. There are many parameters to control in titanium GTAW. Key process parameters are identified as those that can be observed or inferred and those that can be varied to maintain the quality of the process. With such a complex system, involving the integration of many subsystems from different sources, system architecture and data standardisation are critical. An open-architecture CNC controller for intelligent material processing is presented.

Finally, a supervisory interface is proposed that provides a compromise between removing the welder from the immediate task environment; and retaining his abilities to make decisions, handle multiple sensory inputs, cope with poor quality or dubious data and apply judgement. A major part of this is the provision of visual cues to the operator by fusing live images and geometric models of the process.

## 2.     Special Considerations for Welding Titanium

### 2.1     *GTAW for Titanium*

Contrary to widely held opinion, titanium is not a difficult to weld material; but careful preparation is essential and special attention must be paid to procedures to maintain weld quality. A major concern is embrittlement of welds due to contamination. This can be due to oxidation and/or inclusion of foreign material.

GTAW is the most commonly used process for joining titanium. The properties of titanium and titanium alloys are favourable for arc welding as the low densities and high surface tension result in a fluid weld pool. This enables good control of surface profile and weld penetration [4]. Titanium has high reactivity at elevated temperatures and will readily oxidise or combine with other metals. This will embrittle the weld. Oxidation is prevented by shielding heat-affected zones (> 430 °C) with an inert gas such as argon. All surfaces must be clean, free of moisture, dust and grease. Titanium welding is often carried out in specially designated areas that can be kept clear of sources of contamination [5]. Care must be taken to ensure that the electrode does not touch the molten weld pool, or that filler wire does not introduce contaminants.

The requirement for gas shielding often demands the design and manufacture of shielding fixtures to direct gas over heat affected zones. These include [5]:

- Primary fixtures to surround the electrode, arc and weld pool with inert gas.
- Secondary or trailing fixtures to protect the solidifying weld bead.
- Backing fixtures to protect the underside of the weld.

The use of shielding fixtures increases set-up times and reduces welding speeds. The manual welding of titanium is often a two-person task. Mechanised or automatic welding is made difficult because access and path planning is restricted by the fixtures. (Ideally, these considerations should be taken into account when designing components to be fabricated from titanium). An alternative approach is to carry out the welding in a chamber filled with inert gas. This is usually restricted to small components, but a conceptual system has been proposed that allows large structures to be fabricated by a robotic system enclosed in a sealed chamber (Figure 1). Whichever method is used, the use of chambers or fixtures restricts the welder's view and access to the task. The following two sections describe the system architecture.

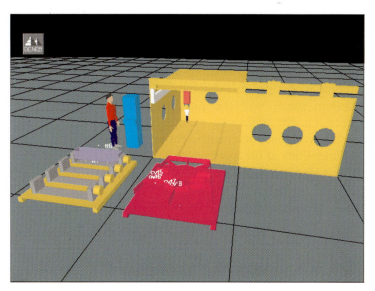

Figure 1 GTAW in a sealed inert chamber.

## 2.2    *Features of an Intelligent Welding System*

When there are large production runs featuring similar weld types, the processes are understood well and therefore predictable. In such cases it is feasible to structure the environment and set-up the process so that 'hard' or dedicated automation can be achieved. This means that there is little variation in workpiece geometry and process parameters are constrained. Many such systems are available commercially. With more complex tasks, it is possible that dedicated machines are only able to perform a small percentage of the welds. In these cases, the flexibility of a programmable and reconfigurable system offers a better solution. Additionally, the complexity of a structure may make the welding process less predictable— e.g. due to heat sinking effects - and so the system may have to adapt its behaviour or process parameters in response to unexpected events or unfamiliar conditions.

The skilled welder is able to control the process based on what can be observed, using his knowledge, experience and dexterity. The welder also has the ability to configure equipment settings for each particular task based on previous experience. The motivation behind the research described here is the need to demonstrate an automatic system that has the features and capability to emulate the performance of the skilled manual titanium welder.

The generic features of such a system (Figure 2) are discussed below. These include the physical attributes, in-process sensing and control, adaptive control, artificial intelligence, task supervision and operator feedback. More detailed explanations of these subsystems can be found in the appropriate sections of this chapter.

The system must have the **physical attributes** to fulfil the task requirements, i.e. the ability to manipulate and move the welding tool relative to the workpiece with sufficient accuracy and speed. The types of weld and dimensions of the workpieces will govern the degrees of freedom required. This research focuses on those tasks that are sufficiently complex to require a multi-axis manipulator.

Changes in the workpiece can result from thermal distortions or misalignment in fixturing. The system must be capable of sensing these conditions and then **adapt** its planned robot **trajectory** to compensate. Seam tracking systems are long established but the surface properties of metals such as titanium pose additional problems.

Welding is a complex multi-parameter process. The skilled welder can vary arc current, using power supply controls, and vary arc length and travel speed dexterously. He will control the process by observing the weld pool geometry. A skilled automatic system will need similar flexibility and control over power supply parameters, trajectory and arc length. The system should sense or infer the condition of the weld pool and come to some form of control decision to alter process parameters. Conventional control schemes are unsuitable because the complex interaction of parameters is too difficult to express in a mathematical expression as part of a control algorithm. Modelling of the process is unrealistic given the fact that many of the parameters are unknown or too difficult to estimate on line. Artificial intelligence techniques, such as fuzzy logic and/or artificial neural networks, are suited to the uncertainties of multiple-input-multiple-output non-linear systems.

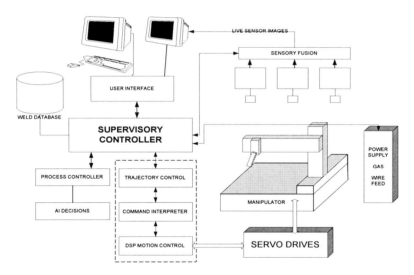

Figure 2    Generic architecture of an intelligent GTAW system.

The manipulator must be run in conjunction with many auxiliary systems such as gas flow regulation, power supply units, wire feeders etc. The sequencing of processes is critical in titanium welding. Supervisory control is needed but the architecture of the system and problems of equipment interfacing require careful consideration.

The subsystems for an automatic system exist and much research effort has been put into the individual areas. There are enough examples of robotic systems working in skilled applications for us to be confident that, given enough time and resources, a system will perform fully automatic titanium welding to required quality standards. There are enough outstanding issues for this goal to be a valuable research exercise. However, the cost and complexity of such a system, combined with the need to prove reliability, will delay the implementation of commercially available fully automatic systems. As an intermediate step, it was proposed to develop a system that can *assist welders, in semi-automatic or tele-operated mode, for applications where the welder cannot get the required access to the workpiece, to manipulate the torch or monitor the process effectively*. This would utilise the sensing and control features of a fully automatic system but retain the operator in a supervisory role, thus allowing a gradual implementation and validation of automation techniques.

### 2.3  *Subsystems and Components*

**Manipulator** – a multi-degree of freedom mechanism for positioning and orientating the welding tool.

**GiCNC** – software for converting weld paths into robot trajectories. These trajectories have to be converted into individual joint commands. These must be coordinated so that the trajectory of the tool on the final link is smooth.

**Servo drives** – amplifiers and motors that respond to control commands to drive the manipulator joints.

**Low-level control** – a motion command is generated in software. This is converted at a low-level hardware interface to 'real' electrical control signals for the servo drives.

**Sensory systems** – There are two important levels of sensory system: low-level control sensors for the servo systems, and process sensors. Process sensors provide information for high-level decision-making so that process or trajectory can be adapted as necessary.

**Vision system** – the vision system is used to monitor progress of the tool against workpiece and can also be used to provide images of the solidified weld bead to a system operator.

**Seam-tracking** – a seam-tracker detects the weld seam ahead of the weld torch and is used to adjust the torch position to correct for alignment errors.

**Weld pool monitoring** – weld pool size and profile are the most important process control variables. Vision systems, infrared sensors and arc sensors have been used to provide information of the weld pool characteristics.

**High-level control** – executive decisions that a skilled welder makes, based on observation of process and experience. To automate these decisions is non-trivial, and usually involves some form of **artificial intelligence** such as neurofuzzy reasoning. Alternatively, the decision-making can be carried out semi-automatically by an operator with a high-level controller converting user commands to system commands.

**Weld database** – information on weld dimensions and procedures that can be converted automatically into trajectory information by task planning functions of the CNC module.

## *Auxiliary Process Equipment*

**Power supply** – the welding power supply is a major influence on the process and weld quality. The electrical parameters must be coordinated with tool trajectory and so the power supply must be controlled by the system. For GTAW of titanium, there are important sequencing and interfacing requirements that must be adopted to protect the weld from embrittlement.

**Gas** – inert gas shielding must be established before arc striking and maintained until the workpiece has cooled below the oxidation temperature.

**Wire feed** – the feed rate of filler wire has to be related to the weld travel speed and welding current, and therefore the CNC controller.

## 3. Manipulator Configuration

### 3.1 *Selection of Axes*

The selection of mechanical configuration and number of axes is governed by the type of welds that must be performed, and what access restrictions there are in the workspace. A machine, which can attain the position and orientation of the weld torch to perform all types of weld on a 3-dimensional structure, requires at least 6 degrees of freedom. A redundant manipulator, having more than 6 d.o.f., may be required to navigate around obstructions; i.e. to overcome degeneracy in the work volume. Having many degrees of freedom is not always desirable as additional axes introduce extra cost, complexity, weight, difficulty in control, reduced rigidity, increased error and are not always necessary.

Looking at a simple case of welding a longitudinal seam (Figure 3a), it is clear that this can be achieved with a single axis machine. A robotic manipulator would not be a good choice if this were the only type of weld, as this solution could be easily provided in 'hard-automation' using a single slideway. If required, additional degrees of freedom can be added without increasing the complexity of the basic machine. For example a carriage can be added to provide lateral oscillation of the welding torch.

Arbitrary lines in a plane can be accomplished with the addition of another degree of freedom (Figure 3b). However, if the orientation of the tool on the plane is important—as it might be when using shielding fixtures and wire feeders—an extra rotary axis will be necessary (Figure 3c).

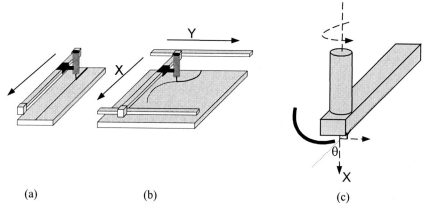

(a)                              (b)                              (c)

Figure 3 Configuration of welding fixturing and positioning devices
(a) single X-axis, (b) 2 d.o.f. with X and Y, (c) fixtures attached to torch
that require orientation by additional rotary axis.

Circumferential welds only require one degree of freedom. In this case the cylindrical workpiece can be rotated relative to a stationary weld tool. Welding lathes facilitate complex paths on cylindrical workpieces and maintain good rigidity.

In general, 3 axes are required to position a tool within a work volume and a 3-axis wrist is needed to provide all possible orientations of the tool. In many cases a two-axis wrist is enough for welding. If required, a missing axis on the robotic manipulator can be compensated for by using a separate device to manipulate the workpiece relative to the tool. This has the advantage of better rigidity than can be obtained in the open kinematic chain of 6-axis manipulator.

## 3.2    *The Experimental and Demonstration Welding Manipulator*

The research projects, associated with the work presented here, are primarily concerned with the welding of box-type structures. A manipulator was required for evaluating welding parameters, sensors and control schemes, as well as demonstrating the feasibility of manufacturing solutions. Referring to the types of welds in Figure 4a, it can be shown that any of the weld paths can be executed with a five-axis configuration. (However, it will be shown that not all of the welds can be performed with the same 5-axis configuration). On Figure 4, welds A and G require 3 d.o.f. for position and one rotary axis for orientation of tool fixtures. Welds B, E and H require an additional rotary axis to orientate the weld tool at 45 degrees for the tee-joint. If weld C is possible with a 5-axis manipulator, welds D and F will not be unless the tool fixture alignment is reconfigured.

As 6-axis coordinated motion is never required in any single weld, it was decided that a 5-axis system (Figure 4b) would provide the best compromise between system capability and cost; as well as being more rigid, less complex, and lighter than a six-axis manipulator.

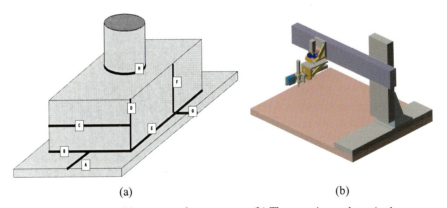

(a)                                              (b)

Figure 4 (a) Weld types on a box structure, (b) The experimental manipulator.

Positioning is achieved by three orthogonal linear axes X, Y and Z (Cartesian configuration). Orientation of the tool is accomplished using two rotary axes, A and B. When producing horizontal welds, the first rotary axis, A, allows control of the orientation of the tool's leading and trailing edges to match the direction of the weld. The second rotary axis, B, allows inclination of the tool to perform different types of weld such as a 45° angle for a tee-joint.

## 3.3    *Machine Kinematics*

The following section presents an analysis of the 5-axis manipulator kinematics. The technique is straightforward and details can be found in most standard robotics textbooks [6, 7]. An understanding of the machine's kinematic behaviour is particularly useful in this application on several counts:

- In remote-system operation, simulation is needed.
- The kinematic analysis is used to derive the algorithms for trajectory planning.
- The kinematic equations are used in image processing of camera views.
- The workspace is confined so the configuration had to be verified at the design stage, and collision avoidance tools are needed for trajectory planning.

The analysis is reproduced here, not only for the reader unfamiliar with robotics, but to illustrate that a methodical approach is necessary for thorough understanding of system capabilities and avoidance of costly errors.

### 3.3.1    *Assigning Coordinate Frames*

In such a system that involves inputs of a multi-disciplinary project team, it is essential that a convention for assigning coordinate frames is adopted and adhered to throughout the life of the project. For this reason, a systematic and well-established approach for defining the relationship between the coordinate frames of the various mechanical links of the manipulator is adopted. (It must be remembered that a simple error in sign, or use of degrees as opposed to radians, can prove to be disastrous when computers are used to control machines). Therefore, all measurements are in mm and radians, all coordinate systems are right-handed, and Denavit-Hartenberg parameters are used to define each frame. The procedure to assign coordinate frames is covered in the following steps.

## a. Define Coordinate Frame of Base $F_0$

The origin of the base frame (world coordinate origin or frame 0) is defined to exist at an easily referenced position on the manipulator worksurface. Calibration checks have to be made with reference to this point so it must be easy to locate and unambiguous. The frame x-axis ($x_0$) is parallel with the first link of the manipulator, the z-axis ($z_0$) is vertically upwards and the y-axis ($y_0$) is chosen to complete a right handed coordinate system.

## b. Identify and Number Links and Joints

These are shown on Figure 5(a). The base is designated link 0 and the robot frames are in the following order – *X, Z, Y, A, B*, tool

## c. Locate z-axes for Each Link

For each rectangular jointed link, the z-axis is along the line of motion of the link. As there are an infinite number of parallel lines that satisfy this criterion, it was decided to define this as being along the centroidal axis of the link's mechanical slideway. For each rotary axis, the z-axis lies along the axis of rotation of the link.

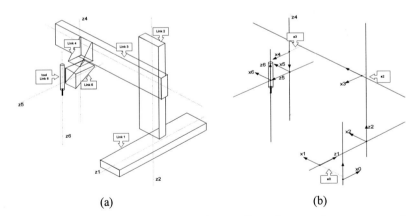

(a)                                      (b)

Figure 5 (a) Manipulator links and z-axes, (b) Coordinate frames and common normals.

## d. Locate Common Normals between Axes

This is the mutually perpendicular line between $z_i$ and $z_{i-1}$ and will define the parameter $a_{i-1}$. When the z-axes intersect, there is no common normal and the parameter $a_{i-1}$ will be zero and the frame origin is located at the intersection (Figure 5b). Only three of the possible seven $a_i$ are non-zero.

## e. Locate Origins of Frames

The origin $O_i$ is located a the intersection of $a_{i-1}$ and $z_i$

## f. Locate x-axis (and y)

The x-axis is chosen to minimise the D-H parameter $\theta$ or make it positive. The y-axis completes a right-handed coordinate system with x and z.

## g. Find D-H Parameters for Each Link

For each link the following can be determined form the physical dimensions of the manipulator.

- The displacement $a_i$ along common normal from $z_i$ to $z_{i+1}$.
- The twist angle $\alpha_i$ about $x_{i+1}$ of $z_i$ to $z_{i+1}$.
- The offset $d_i$ along $z_i$ from $x_i$ to $x_{i+1}$.
- The angle $\theta_i$ about $z_i$ of $x_i$ to $x_{i+1}$.

Table 1 D-H parameters for links 0 to 5 and tool.

|          | 0        | 1        | 2         | 3         | 4              | 5        | t        |
|----------|----------|----------|-----------|-----------|----------------|----------|----------|
| $\alpha$ | $+\pi/2$ | $-\pi/2$ | $+\pi/2$  | $+\pi/2$  | $+\pi/2$       | $+\pi/2$ | 0        |
| $a$      | 123      | 0        | 89        | 150       | 0              | 0        | 0        |
| $\theta$ | $+\pi/2$ | 0        | $+\pi/2$  | 0         | $+\pi/2 +dA$   | $dB$     | $-\pi/2$ |
| $d$      | 37       | $xh + dX$ | $zh + dZ$ | $yh + dY$ | 200            | 100      | $-t$     |

Note:
$xh$, $yh$ and $zh$ are the home positions of the axes.
$dX$, $dY$, $dZ$, $dA$, and $dB$ are the joint commands for he axes.

All coordinates are expressed in homogeneous form (Equation (1a)). The relationship between one coordinate frame $F_i$ and the preceding $F_{i-1}$ is given by the transform matrix $T_{i-1}^{i}$. This is found from the D-H parameters for each link using Equation (1b).

$$\mathbf{p} = \begin{bmatrix} p_x & p_y & p_z & 1 \end{bmatrix}^T \tag{1a}$$

$$T_{i-1}^i = \begin{bmatrix} c\theta_i & -s\theta_i & 0 & a_i \\ c\alpha_i s\theta_i & c\alpha_i c\theta_i & -s\alpha_i & -d_i s\alpha_i \\ s\alpha_i s\theta & c\alpha_i c\theta_i & c\alpha_i & d_i c\alpha_i \\ 0 & 0 & 0 & 1 \end{bmatrix} \tag{1b}$$

As examples, consider the frame attached to link 3 which is attached to link 2 by a prismatic or sliding joint, and link 5, which is attached to link 4 by a rotary joint. From Equation (1b), the transformation matrices to relate these frames to the preceding frame are found to be:

$$T_2^3 = \begin{bmatrix} 0 & 0 & 1 & 0 \\ 1 & 0 & 0 & 89 \\ 0 & 1 & 0 & zh+dZ \\ 0 & 0 & 0 & 1 \end{bmatrix} \quad T_4^5 = \begin{bmatrix} -s(dA) & 0 & c(dA) & 0 \\ c(dA) & 0 & s(dA) & 0 \\ 0 & 1 & 0 & 200 \\ 0 & 0 & 0 & 1 \end{bmatrix}$$

In most cases, we wish to express coordinates with respect to the base frame. To transform a coordinate, we progress down the kinematic chain using each transform in sequence until we reach frame 0; e.g. to find the coordinates of a point p on link 5 in frame 0 coordinates:

$$p_0 = T_0^1 T_1^2 T_2^3 T_3^4 T_4^5 p_5 = T_0^5 p_5 \tag{2}$$

The values of the elements in $T_0^5$ will depend on the instantaneous joint values as well as all the physical offsets in the kinematic chain from frame 0 to frame 5.

### 3.3.2    Kinematic Simulation

Once the frame transformation matrices have been established, it is easy to determine the position and orientation of any component attached to any of the rigid links for any joint position. A useful application of this is in graphical simulation of the manipulator's motion. Figure 6 is the graphical output of a simulation in Matlab. This 3D plot is produced by passing an array of joint coordinates to a user-defined function, **weldpos()**. Solid boxes, defined by the coordinates of their 8 vertices, represent the links of the manipulator. Each box is defined in its own coordinate frame and can be scaled to match the dimensions of the physical link. The procedure for drawing a manipulator for a given joint command is as follows.

Starting at the first link:

- Determine the transformation matrix from the link's frame to the preceding frame using the link parameters (Table 1) and the joint variable for that link.
- Determine the transformation matrix from the link to the base frame by matrix composition as in Equation (2).
- Find the coordinates of the vertices of the box in frame 0 by premultiplying their values by the link to base frame transformation matrix.
- Draw the box by plotting each face (or rectangular patch) as a 3D plot
- Repeat for the next link along the kinematic chain.

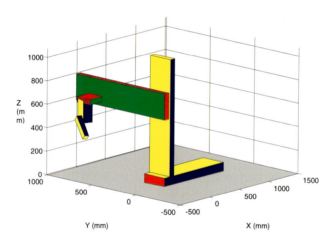

Figure 6 Kinematic simulation in Matlab.

The function **weldtest** takes five arguments, the five joint angles (*dX*, *dY*, *dZ*, *dA*, *dB*), draws the manipulator and returns the transformation matrix (from tool to base frame) to the Matlab workspace; e.g. to produce Figure 6.

$Tt0$ = **weldtest**(-220,100,80,$\pi/7$,$\pi/4$)

These software tools provide a very convenient method of testing algorithms and visualising the behaviour of the manipulator. As the programming environment is interpretative and requires no compilation, the kinematics can be tested without having to spend time debugging C/C++ code. It also has the advantage of being easy to integrate with other tools such as Simulink. A limitation of numerical simulation is that we have to perform many simulations to see the effect of varying different parameters. A deeper understanding of the machine's behaviour can be acquired by analysing the transformations symbolically. Equation (3) shows the matrix composition of the tool transformation in terms of all the D-H parameter symbols.

$$
\begin{bmatrix} 0 & 0 & 1 & 0 \\ 1 & 0 & 0 & a0 \\ 0 & 1 & 0 & d0 \\ 0 & 0 & 0 & 1 \end{bmatrix}
\begin{bmatrix} 1 & 0 & 0 & 0 \\ 0 & 0 & 1 & 0 \\ 0 & -1 & 0 & xh+dX \\ 0 & 0 & 0 & 1 \end{bmatrix}
\begin{bmatrix} 0 & 0 & 1 & 0 \\ 1 & 0 & 0 & a2 \\ 0 & 1 & 0 & zh+dZ \\ 0 & 0 & 0 & 1 \end{bmatrix}
\begin{bmatrix} 1 & 0 & 0 & a3 \\ 0 & 0 & -1 & 0 \\ 0 & 1 & 0 & yh+dY \\ 0 & 0 & 0 & 1 \end{bmatrix} \cdot
$$

$$
\begin{bmatrix} -s(dA) & 0 & c(dA) & 0 \\ c(dA) & 0 & s(dA) & 0 \\ 0 & 0 & 0 & d4 \\ 0 & 0 & 0 & 0 \end{bmatrix}
\begin{bmatrix} c(dB) & 0 & s(dB) & 0 \\ s(dB) & 0 & -c(dB) & 0 \\ 0 & 0 & 0 & d5 \\ 0 & 0 & 0 & 1 \end{bmatrix}
\begin{bmatrix} 0 & 1 & 0 & 0 \\ -1 & 0 & 0 & 0 \\ 0 & 0 & 1 & -t \\ 0 & 0 & 0 & 1 \end{bmatrix}
$$

$$= $$

$$
\begin{bmatrix}
c(dA) & s(dA)\cdot c(dB) & s(dA)\cdot s(dB) & -s(dA)\cdot s(dB)\cdot t - c(dA)\cdot d5 - a3 - a2 + xh + dX \\
-s(dA) & c(dA)\cdot c(dB) & c(dA)\cdot s(dB) & -c(dA)\cdot s(dB)\cdot t + s(dA)\cdot d5 + yh + dY + a0 \\
0 & -s(dB) & c(dB) & -c(dB)\cdot t - d4 + zh + dZ + d0 \\
0 & 0 & 0 & 1
\end{bmatrix}
$$

$$(3)$$

## Interpreting Tool Transformation Matrix

From the matrix in Equation (3), the consequence of altering any of the physical parameters can be deduced. We can see how each system parameter influences the overall transformation matrix for tool to base frame coordinate system. The 3×3 submatrix in the top left corner is the frame rotation matrix. The columns of this are the projections of the tool's $x$, $y$ and $z$ axes on the base frame. The fourth column gives the coordinates of the tool frame origin in frame 0 coordinates.

Closer inspection of the first column of the rotation matrix shows us that the tool's x-axis has components in $x_0$ and $y_0$ (governed by joint angle $A$) but has no component in $z_0$; i.e. no vertical component. This means that if the tool is aligned so that its direction of travel points along this axis, the tool cannot pitch. This limitation is due to the manipulator only having 5 degrees of freedom. In practice, if this type of motion is required for a weld, this limitation is overcome by physically reconfiguring the tool so that its direction of travel is along its y-axis; i.e. unclamping the tool and fixtures and rotating by $\pi/2$ radians. It is important to note that the manipulator is effectively a different machine (in terms of its kinematics) if its mechanical configuration is altered in this manner, and modified kinematic equations would have to be used. The machine would still be limited in orientation but the limitation is now in motion normal to the weld direction.

### 3.3.3    Inverse Kinematics

We have seen how joint angles can be used in forward kinematic transformations to find Cartesian coordinates. We now need a method for determining the joint commands to achieve a specified Cartesian position and tool tip orientation.

As the origin of the tool frame was defined to be at the tool tip, its position and orientation can be taken directly from the tool transformation matrix. More generally, to find the position of any point attached rigidly to the tool frame, pre-multiply its position vector by the tool transformation matrix, Equation (3).

The link displacement parameters and joint home positions ($a_i$, $d_i$, xh, yh and zh) have been replaced by actual values in Equation (4). This gives the tool tip position vector in frame 0 coordinates as a function of the five joint angles.

$$p_t = \begin{bmatrix} 0 & 0 & 0 & 1 \end{bmatrix}^T$$

$$\begin{bmatrix} x \\ y \\ z \\ 1 \end{bmatrix}_0 = T_0^t p_t = T_0^t \begin{bmatrix} 0 \\ 0 \\ 0 \\ 1 \end{bmatrix} = \begin{bmatrix} 200s(dA)s(dB) - 100c(dA) + 101 + dX \\ 200c(dA)s(dB) + 100s(dA) + 683 + dY \\ 200c(dB) + 397 + dZ \\ 1 \end{bmatrix} \qquad (4)$$

To solve the inverse kinematics for a location along a welding trajectory:

- The first step is to set the joint angle B. This is determined by the type of weld (either horizontal, angled or vertical), and is usually fixed for the duration of that weld pass.
- The joint angle $A$ is determined next. This value is important if there are shielding fixtures, wire feeders or sensors attached to the tool that need to be aligned with the weld direction. This angle is found by aligning the tangent of the weld path with x-axis of the tool frame, $x_t$.
- With the two rotary joint values determined for a path point, the linear joint displacements $dX$, $dY$ and $dZ$ are solved by rearranging Equation (4) to give Equation (5):

$$dX = x_0 - 200s(dA)s(dB) + 100c(dA) - 101 \tag{5a}$$

$$dY = y_0 - 200c(dA)s(dB) - 100s(dA) - 683 \tag{5b}$$

$$dZ = z_0 - 200c(dB) - 397 \tag{5c}$$

It can be seen that, due to the simple configuration of the manipulator, the equations for tool position are also simple. It could be argued that these equations can be derived from inspection and 3D geometry without the need to use coordinate frames for each link. However, the danger of producing control code directly from geometry means that many offsets are hidden within the code and, if they are changed due to physical modification, it becomes difficult to track which code needs to be modified. By using transformations, only the code that defines the transformation needs to be updated. The matrices provide natural data objects that can be protected within large applications.

### 3.3.4 Trajectory by Decomposition of Tool Transformation Matrix

If a trajectory is available as vector coordinates – as it might be from the output of another software package – the inverse kinematics are easily solved by decomposition of the tool transformation, Equation (3). Consider the case of welding a tee-joint in a circular arc (Figure 7) at a point where the vector defining the direction of the weld is defined by the unit vector $[0.9010, -0.4339, 0]^T$ and a normal to this defined by the unit vector $[0.6371, 0.3068, 0.7071]^T$.

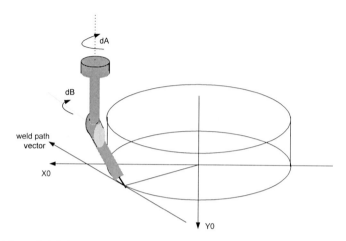

Figure 7 Solving manipulator inverse kinematics on arc trajectory.

The direction vector corresponds to the tool x-axis and the normal corresponds to the tool z-axis. As the coordinate frame is orthonormal the $y$ vector is equal to the cross product of $z$ and $x$.

$$y = z \times x \tag{6}$$

The rotation matrix of the tool transform is then

$$R_0^t = \begin{bmatrix} x & y & z \end{bmatrix} = \begin{bmatrix} 0.9010 & 0.3068 & 0.3038 \\ -0.4339 & 0.6371 & 0.6371 \\ 0 & -0.7071 & 0.7071 \end{bmatrix}$$

$$= \begin{bmatrix} c(dA) & s(dA)c(dB) & s(dA)s(dB) \\ -s(dA) & c(dA)c(dB) & c(dA)s(dB) \\ 0 & -s(dB) & c(dB) \end{bmatrix} \tag{7}$$

As a first check, element $R_{31}$ must be zero. Joint angle $dA$ is solved from

$$dA = \text{atan2}\,(-R_{21}, R_{11}) = 0.448\,\text{rad} \tag{8}$$

Joint angle $dB$ is solved from

$$dB = \text{atan2}\,(-R_{32}, R_{33}) = 0.7854\,\text{rad} \tag{9}$$

## 4.     Process Control

In the fabrication of high value components from titanium, maintenance of weld quality is of utmost importance. Furthermore, rework is undesirable and costly. It is therefore imperative that correct procedures are identified and adhered to. The need for good preparation and protection from contamination has already been stressed.

Arc welding is a process involving many interrelated phenomena and, even in a highly structured environment, it will be difficult to constrain all the variables so that the process behaves in a predictable or repeatable manner. Therefore, the process must be monitored so that any deviations can be detected and corrected before the weld ceases to conform to required quality criteria.

There are two major concerns:

- Is the weld in the correct place? (Position of weld pool and seam)?
- Does the weld bead have the correct profile? (Temperature and size of weld pool)?

Incorrect weld position can be caused by misalignment of the weld tool due to robot errors, thermal distortion of the workpiece or misalignment of the workpiece in fixtures. Of these, thermal distortion is the hardest to predict although its effects can be reduced through appropriate choice of fixturing, lengths of weld runs and power supply parameters. If misalignment cannot be eliminated to conform to the specified tolerance through careful set-up, the robot trajectory can be adapted with the use of seam-tracking control. There are three major types; mechanical, through-the-arc and optical seam-tracking.

The quality of a weld can be assessed according to the penetration in the base metal, bead width, weld reinforcement or height, weld surface appearance and weld metal deposit. These factors are all influenced by heat input.

Expressed simply, in weld process control "the aim should be to achieve a good bead shape with minimum heat input" [4]. Too much heat will cause problems with thermal distortion and oxidation of heat affected zones. Too little heat will give insufficient penetration and bead size. The rate of heat input is governed by the arc voltage and current. How much electrical power to apply is influenced by how heat energy is dissipated by the workpiece and fixturing. Clearly, when faster weld traverse speeds are employed, greater power input is needed from the welding power supply.

The primary process parameters are:

- Arc current.
- Arc voltage.
- Welding speed.
- Pulse profile.
- Up-slope or down-slope welding.

The following are secondary parameters. They have an effect on the quality but they are usually set according to other factors, and so would not be used to directly control the weld pool.

- Shielding gas composition and flow rate.
- Filler wire diameter and feed rate.
- Nozzle diameter.
- Electrode diameter and end angle.
- Number of passes.
- Time before cooling.
- Length and sequence of welds (to reduce distortion).
- Fixturing and backing plate design.

## 4.1    *Critical Process Parameters*

Current and welding speed are the two main parameters that can be varied to control the weld pool. In manual GTAW, the current/speed setting is usually set according to the operator's preference.

### *Power Supply*

The welding power supply (Figure 8) is critical to the characteristics of the weld. The following parameters are set on a typical pulsed-current GTAW power supply unit.

- High-pulse current.
- Low-pulse current.
- Pulse frequency.

• Pulse duty cycle.

The pulse-current supply offers improved penetration due to the localisation of intense heat input for a restricted length of time [8]. Fusion occurs during the high-current pulse, and partial solidification during the low-current pulse. This gives the weld the appearance of a series of overlapping spots. The average current is reduced, the width of the weld is smaller, less metal needs to be melted and therefore welding speeds can be increased.

The power supply used in a robotic system must have an interface that allows its parameters to be controlled by the system controller's software. An increasing number of supplies are available with interfaces that are suitable for direct interfacing, or can be easily adapted to take either digital or analogue control signals. The power supply used in this research was a Fronius Transtig 2000. The Liburdi LTP100 power supply facilitates weld sequence programming via RS-232 connection to a host computer, as well remote control and telemetry via an analogue and digital interface. The analogue input controls arc current. The telemetry allows monitoring of current and voltage, and provides a digital signal to indicate when the arc is established.

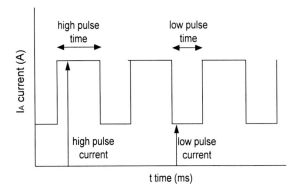

Figure 8 Pulse welding profile and controls.

Arc voltage is primarily a function of the arc-gap distance. This is usually set depending on the electrode type and dimensions, in accordance with weld procedure. This would not normally be used as a control parameter once a weld pass had been started.

## 4.2    *Inert Gas Shielding*

Argon is usually the preferred choice of shielding gas as it gives better arc stability. If a trailing shield is used, its size will determine maximum weld speed because the shield must protect the solidifying weld bead until the weld temperature falls below 430 °C. The gas flow should be uniform and non-turbulent.

Other shielding considerations that must be taken into account when integrating the process with the system controller include:

- Separately controlled gas supplies are recommended for the three shielding zones: primary, secondary and backing.
- The gas flow should be started before the arc to pre-purge the shields of all air.
- Gas flow should be maintained after extinguishing the arc until the metal has cooled below the oxidation temperature.

If a shielding chamber is used, there will be many other environmental control issues introduced; such as evacuation and purging of air, monitoring of gas composition, heat dissipation, etc. These issues are beyond the scope of this current discussion

## 4.3    *Other Procedures*

- If problems with moisture are anticipated due to high humidity or low temperatures, preheating may be required.
- Filler wire should be fed in smoothly into the weld zone. Intermittent dipping is inadvisable as this may cause turbulence and contamination of the hot end of the wire as it leaves the protection of the shielding gas.
- Joints should be adequately tacked to prevent relative movement between the parts during welding.
- The welding sequence and technique should aim to balance the thermally induced stresses around the neutral axis of the component.
- Heat build up from previous weld runs can lead to surface contamination on subsequent passes. In some instances it may be necessary to leave the work to cool before further welding is carried out. Another approach is to make long welds in shorter sections. In

addition to helping with cooling, sequence welding can also be effective in reducing distortion [4].

## 4.4 *Sensing and Monitoring*

It has been stated that the surface properties of the weld pool are the main process parameters to control. Manual operators observe weld pools and changes in their-pool geometry, and decide how best to change the welding input parameters; e.g., arc current, pulse-profile or traverse speed. A reliable sensing method is needed to evaluate these properties to provide the input for a control scheme. Images acquired with laser strobe systems [9] have been used to assess weld pool geometry [10, 11, 12]. Some limitations of this equipment are that it is expensive, delicate, and intrusive. It is therefore not suitable for many industrial applications. Passive devices rely on the arc to provide illumination of the weld pool. Richardson and Gutow [13] conceived a device incorporated into the welding torch, thus overcoming the problems of sensor access in confined spaces. The imaging optics are co-axial with the electrode, blocking out most of the intense glare from the arc. This concept has been used by several researchers to study weld pool characteristics [14, 15]. Weldware offers this concept as a product that can show weld joint, weld pool and solidified weld bead.

Depth of penetration is a critical weld quality parameter. Various schemes have been investigated for estimating this [16]:

- Looking at intensity of the light coming from the backside of the weld, and correlating light intensity with penetration.
- Infrared equipment has been used to infer penetration depth from weld pool temperature profiles [17].
- Sensing using acoustic emission [18].
- Radiography.
- Extensive research has been carried out on techniques using the resonant frequency of weld pool oscillations to infer geometric properties. Pool oscillation was measured using arc voltage and arc light fluctuations.

TIGpc is a weld penetration system from Weldware, which uses digital signal processing of arc voltage or light intensity, to detect weld pool oscillation frequencies.

A number of suppliers offer systems that analyse sensory data and use this to detect known fault conditions. Impact Engineering provides a monitoring system that analyses arc current, supply voltage, shielding gas flow, and wire feed speed to verify process parameters [19]. When problems are detected, diagnostic information is displayed for the operator to take action. Other systems have correlated weld bead porosity and joint misalignment with the weld signature (voltage and current). Quinn et al [20] used current and voltage signals to detect welding defects in pulsed current GMAW. They tested on faults such as lack of shielding gas, oily parts and melt through. Their system did not detect off-joint welding.

Seam-tracking systems are common and many options are available commercially. Mechanical devices use a stylus to follow the seam. These can encounter difficulties when there are obstructions such as tack welds or when the joint is tight. Through-the-arc sensors provide a neat solution as they require no extra hardware to be mounted on the weld tool. These systems infer position from the arc voltage (and current). As the arc voltage is related to the arc length, an oscillating (or weaving) torch will give an arc voltage that also oscillates. Although subject to much noise and other parameters, the signal is good enough to infer the centre of the vee-groove.

Laser devices use triangulation to determine depth information from a scanning laser stripe [21, 22]. These have been successfully applied in many applications but problems have been reported when dealing with titanium due to its high reflectivity causing multi-path errors. They do have the advantages of being able to work at higher speeds, cope with arbitrary shaped grooves, and do not require the torch to weave.

## Other Sensing Options

Trends in weld quality can be detected from images of the solidified weld bead. These can be inspected for size, shape and uniformity. Colour images can be used to detect for oxidation due to air contamination. Layers of oxide generate interference colours on the metal surface. As the oxidation increases, the colours change and thus give a warning of an impending or existing problem.

An image of the electrode (once the arc is extinguished), showing signs of oxidation, is indicative of air contamination of the primary gas shield.

## 4.5    Schemes for Process Control

Although welding is a multi-variable process, a control scheme should be simple. Testing to establish optimum weld parameters is highly advised. In this way many of the parameters such as, electrode size, arc length, pulse profile, gas flow, etc., can be set-up beforehand. The process control scheme can then concentrate on the critical parameters that have most influence on the quality of the process. Gintic's intelligent welding test system control scheme is based on:

- Adjusting position of torch relative to seam.
- Adjusting arc current and/or weld traverse speed.

The position loop is closed using sensory information from a seam-tracking device. This is relatively straightforward as the positional error signal is readily available from the sensor processing algorithm. The error can be used to add an offset to the original weld path trajectory. The processing algorithms must be robust to noise and spurious signals, otherwise the weld path will be subjected to erratic lateral movements.

The process control loop can be closed with a sensory system based on weld pool inference. A vision-based technique has been demonstrated at Gintic [12]. Weld pool oscillation techniques show good promise as they are less intrusive if they are based on arc voltage analysis. Once the oscillation frequency is detected, the welding current and/or travel speed can be adjusted to maintain the frequency in a desired range and thus control penetration. Whatever sensing method is used, the control algorithm must be robust to cope with the range of phenomena encountered in GTAW.

## 4.6    Use of AI in Automatic Welding

The complex nature of GTAW makes it a suitable candidate for AI techniques such as fuzzy inference or artificial neural networks (ANN). The welding process has many input parameters and many output parameters. It has been stated that conventional control is troublesome because it is difficult to find mathematical expressions to cope with this level of complexity; i.e. multiple inputs, non-linearities, time variance, partly known systems, etc. Incorporating knowledge, automated knowledge extraction and learning are key issues for the application of more advanced, intelligent control schemes.

Fuzzy control systems are suited to these conditions. A fuzzy system can be thought of as an arbitrary mapping between inputs and outputs or a 'black box' solution. The advantages are:

- Tolerance of imprecise data.
- Rules are formatted using natural language descriptions and it is easier to encapsulate skilled operators' expertise and judgement.
- Multiple inputs and multiple outputs.
- Non-linear behaviour can be incorporated.
- Welding parameters can be adjusted or tuned.

Artificial neural networks have proved successful in applications such as pattern recognition, identification, machine vision and process control. ANN systems are tolerant to noise, can account for non-linear behaviour, and are suitable for multiple-input-multiple-output (MIMO) non-linear systems. ANN's were used for analysis of weld pool images at Gintic to extract geometric properties, such as the width and length, perimeter, and area of the weld pool.

Another major advantage of ANNs is their ability to be trained. If there is sufficient training data available, the network will learn to associate inputs with outputs. Combining fuzzy inference systems with neural networks leads to another variation – adaptive neurofuzzy systems. These are fuzzy systems that can learn; i.e. their membership sets are tuned using neural network training algorithms.

Given the advantages described above, it is not surprising that many researchers have investigated the applications of fuzzy and neural systems in arc welding. [23, 24] compared conventional PID with fuzzy control and single neuron self-learning proportional sum and differential (PSD) control schemes. The system is based around a vision sensor that monitors both topside and backside of the weld with the aim of controlling backside width of the weld in pulsed GTAW. The results produced comparable performance for fuzzy control and PID but marked improvement with self-learning PSD. Zhang et al [25] uses neurofuzzy systems to monitor the topside of weld, estimate topside and back –side bead widths and hence control weld current and weld speed.

## 5. CNC and Low-Level Control

The complexity of automation systems for skilled tasks demands many subsystems and distribution amongst many different processes. As new or improved products become available, the system will require reconfiguration. Alternatively, it may be necessary to remove certain features due to maintenance or process change requirements. The system architecture must be open and organised in such a way that it is not dependant on specific hardware.

It should be possible to issue executive level commands without direct interaction with low-level protocols or commands. This type of structure is a layered hierarchy. In such a structure, supervisory goals are decomposed into increasingly primitive subtasks until these subtasks are directly applied to mechanical or electrical components [26]. For example, a high-level command is descriptive of a weld type. At the next level this is converted into dimensions and other process specific data using the database of weld procedures. These dimensions are mapped onto robot workspace coordinates.

A further level down introduces hardware specific functions, as the weld-path coordinates are used to generate joint space trajectories based on the manipulator's inverse kinematics. Next, this trajectory needs interpreting and passing in a form for low-level functions of the motion controller. This is now at the hardware level as the controller produces electronic control signals (digital or analogue) for the servo amplifiers. The amplifiers then output the electrical power to drive the joint actuators.

By layering in such a hierarchy, subsystem components can be exchanged without affecting layers more than one level of abstraction away. For example, if the motion controller is exchanged with that of a different manufacturer, it may only be necessary to replace the software library that converts between joint commands and motion control hardware specific commands.

### *CNC Programming (GiCNC)*

GiCNC is a hierarchical open-architecture CNC controller developed at Gintic for intelligent material processing (Figure 9). The hierarchical modularisation makes it easy to be scaled down or scaled up to control more axes for coordinated motion. With the open architecture, advanced network technologies are readily available and network features are accessible to the CNC systems.

Using object-oriented technology, the modular software is highly portable and scalable. The software is PC based, a platform which is gaining popularity and acceptance in industrial control.

Until standards for exchange of welding procedures and equipment interfacing can be agreed and implemented, there will still be considerable cost associated with software and system integration. These costs can be two to three times the cost of the manipulator [27]. In 1997 NIST proposed an Advanced Welding Manufacturing System as a testbed for technology and integration issues for intelligent automated arc welding [28]. Associated with this work was the formation of standards for data exchange between computer-based welding systems such as computer-aided design, weld design, welding process planning, real-time process control and quality control.

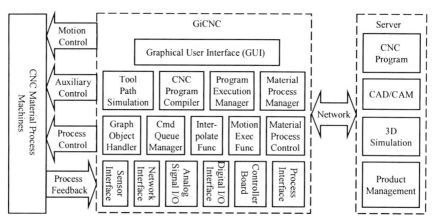

Figure 9 GiCNC – Open architecture CNC system for intelligent material processing.

The American Welding Society established a subcommittee on the *Exchange of Welding Information between Intelligent Systems*. This committee proposed an activity model that describes the information necessary for arc welding without describing the communication mechanisms or encoding. The systems covered include welding process databases, off-line planning systems, CAD systems, system controllers, robot controllers, welding controllers and intelligent sensors. A draft standard, from the A9 subcommittee, for welding cell intra-cell network communication was scheduled to be tested in 2003 as part of the *Intelligent*

*Open Architecture Control of Manufacturing Systems* programme at the Manufacturing Engineering Laboratory, NIST.

## 6. Interactive and Tele-Operated Welding System

Use of computer vision in GTAW was discussed in its role as part of the automatic process control loop. There are still many problems that need to be overcome, before such systems are accepted as reliable and practical enough for use in fully automated systems. An alternative approach is to retain the human operator in the control loop. The operator acts as an observer, checking for correct operation of the automatic systems or making decisions based on measured and presented parameters. The operator is used to do what humans do best and is difficult to automate; i.e. interpreting complex multi-sensory inputs, coping with dubious data, applying judgement and experience. The operator can supervise the process without having to concentrate on skilled manipulation of the torch or having to get close to the process to make observations.

This mixture of automation and human intervention is justified in applications such as this where:

- There is a need to separate task from human due to safety or accessibility concerns.
- The process is complex with many parameters to account for and the operator requires computer assistance, but computer control requires human judgement.
- Parts are of high value, and it is essential to minimise risk of errors that will result in scrap or rework.
- It is difficult to find skilled labour to perform the tasks manually.

## 6.1 *Visual Monitoring*

Visual monitoring of arc welding is difficult due to the intensity of the arc. The human eye is easily damaged by the very bright light, and welding goggles must be used to protect the eye from UV emissions when viewing the arc and weld pool. The arc intensity also creates problems for camera equipment. The intensity can be reduced by appropriate setting of the lens aperture and electronic shutter speed, but there is a high contrast between arc and background as demonstrated in Figure 10. Very little detail can be discerned in either foreground or background. Another problem is that certain pixels become saturated with charge that spills into neighbouring pixels, resulting in blooming of the image. Background information is important in this application so that the operator can: check progression of the weld relative to the workpiece, check for potential collisions, and see if the workpiece has become misaligned.

The intensity problem can be alleviated through the use of optical filters. A combination of neutral density (index 3) and narrow passband filters (660 nm) improves details of the arc and weld pool, but details of background and colour are lost. Jetline offers a colour CCD monitoring system that uses a camera with fast shutter speeds (1/50 000 sec) and spot filters to reduce the intensity of the arc.

Figure 10 CCD camera image of GTAW.

Logarithmic cameras use CMOS circuitry that give a very high dynamic range (120 dB), and can therefore cope with high contrast images. These also have the benefit of random addressing of pixels, which allows faster frame rates and selection of regions of interest (ROI). Because the output is an instantaneous measure of photodiode current, they avoid the problems of blooming associated with charge accumulation and transfer in CCD technology. Logarithmic sensors such as the FUGA 15d are available in RGB and BW versions.

One aim of this research is to use off-the-shelf equipment with a minimum of additional components and configuration. Even though the image is not ideal, it is enhanced using modelling techniques to reconstruct corrupted or occluded information. This gives the operator enough references to interpret images quickly. This is feasible in this application because the environment is structured, workpiece geometries are simple and, as part of an integrated system, the geometric data is available from a database of welding operations.

## 6.2    *Computer Enhanced Vision*

Modelling techniques have been used extensively in virtual reality and simulation for training of operators before undertaking tasks. They are useful in applications where the operator cannot gain access to the task, such as in nuclear reactors, underwater or space exploration. There are, however, not many applications where the environment is known with sufficient certainty to rely on modelling alone.

The welding application is an example where the environment is well defined but uncertainty is introduced by thermal distortion, alignment of fixtures and spurious sensory signals. The workpiece and manipulator can be modelled from CAD data and kinematics. We can also model the camera to visualise how the workpiece and tool will appear on a monitor or the output of a frame grabber. Through integration of these models, the simulated view of workpiece and weld tool is drawn as an ovelay on the live video image for operator verification and monitoring. This is beneficial because:

- The operator is given added confidence in his interpretation of the live image.
- If there is a distinct mismatch between simulation and observed view, this is indicative of some system or process error, and the operator can take action.

### 6.2.1    *Modelling of Camera Views*

The CCD view is modelled as a simple pinhole camera. The camera has its own coordinate frame $F_c$ with the z-axis running along the optical axis. To model a camera view, we need to establish how three-dimensional objects map onto the small two-dimensional CCD image plane. We need to find a transform that converts from the world (or base frame system) $F_0$ to the camera coordinate system, and then a perspective transform that will represent the action of the lens projecting these objects onto the image plane.

The camera coordinate frame can be treated in the same manner as the robot link frames in the kinematic analysis; i.e. it is subjected to translations and rotations away from the base frame. The camera has six degrees of freedom – 3 translations ($x$, $y$ and $z$) plus 3 rotations.

The camera coordinate transform is established using the following procedure:

- Find the displacement transform $T_d$ from Equation (10a), where $X_c$, $Y_c$ and $Z_c$ are the coordinates of the camera's principal point in frame 0.
- The pan angle $\alpha$ is a rotation about the z-axis and is equivalent to the transformation matrix Equation (10b).
- The tilt angle $\beta$ is a rotation about the y-axis, Equation (10c).
- The roll angle $\gamma$ is another rotation about the z-axis, Equation (10d). These four matrices will transform from base frame coordinates to the camera coordinate frame. Other combinations of rotation matrices are also possible—for example roll, pitch and yaw about $X$, $Y$ and $Z$ [29, 30].
- A perspective transformation $T_p$, Equation (10e), accounts for how the lens creates the image on the CCD plane, where $f$ is the focal length.

$$T_d = \begin{bmatrix} 1 & 0 & 0 & -X_c \\ 0 & 1 & 0 & -Y_c \\ 0 & 0 & 1 & -Z_c \\ 0 & 0 & 0 & 1 \end{bmatrix} \tag{10a}$$

$$T_\alpha = \begin{bmatrix} c\alpha & s\alpha & 0 & 0 \\ -s\alpha & c\alpha & 0 & 0 \\ 0 & 0 & 1 & 0 \\ 0 & 0 & 0 & 1 \end{bmatrix} \tag{10b}$$

$$T_\beta = \begin{bmatrix} c\beta & 0 & -s\beta & 0 \\ 0 & 1 & 0 & 0 \\ s\beta & 0 & c\beta & 0 \\ 0 & 0 & 0 & 1 \end{bmatrix} \tag{10c}$$

$$T_\gamma = \begin{bmatrix} c\gamma & s\gamma & 0 & 0 \\ -s\gamma & c\gamma & 0 & 0 \\ 0 & 0 & 1 & 0 \\ 0 & 0 & 0 & 1 \end{bmatrix} \tag{10d}$$

$$T_p = \begin{bmatrix} 1 & 0 & 0 & 0 \\ 0 & 1 & 0 & 0 \\ 0 & 0 & 0 & 0 \\ 0 & 0 & 1/f & 0 \end{bmatrix} \tag{10e}$$

Composing these together in the following sequence will provide the base frame to camera image coordinate transformation, $T_c^0$

$$T_c^0 = T_p T_\gamma T_\beta T_\alpha T_d \tag{11}$$

The projection of any three dimensional coordinate on the camera image plane can be found from

$$p_c = T_c^0 p_0 \tag{12}$$

where

$$p_c = \begin{bmatrix} p_x & p_y & p_z & w \end{bmatrix}^T \tag{13}$$

The x and y coordinates on the CCD image are then

$$x = \frac{p_x}{w} \text{ and } y = \frac{p_y}{w}$$

## 6.2.2    Modelling of Solid Objects

The significant objects in the welding camera views are represented as simple geometric shapes:

- Rectangular plates.
- Chamfered edges.

- Cylindrical welding torches.

Objects can be described by surfaces and edges. These descriptions correspond to how image processing software identifies the objects. Drawing of these objects is straightforward with low computational overhead. Each object is composed of rectangular surface patches; each having four 3D vertices. The objects are drawn in wire-frame or as filled patches. For each view, the data for each object is transformed, using the current view transform, $T_C^0$, to the camera's coordinate frame. Valid data (in front of the camera) is drawn within a screen view scaled and clipped to the dimensions of the CCD device.

All algorithms were tested in Matlab and MathCad before implementation in C++. Figure 11 illustrates a perspective view of a welding torch over a vee-groove joint. Data has been converted to camera data and then drawn as a simple 2D *xy*-plot in Matlab.

Figure 11 Perspective modelling in Matlab.

### 6.2.3    *Implementation in Simulation Software*

The camera view models were incorporated in simulation software to:

- Test positioning of cameras within workspace.
- Determine optimum lens and camera parameters.
- Create virtual reality motion over workpiece to simulate tool motion.

- Present hand-eye viewpoint based on manipulator kinematics.

Static viewpoints are direct C++ implementations of the algorithms developed in Matlab. The user is able to select 6 camera location parameters, lens size and CCD format. Standard pre-configured views are plan, side and end-projection.

In moving views, camera view parameters are updated according to a predefined trajectory at regular time intervals. The first type of animation is for a moving camera which allows the task planner to view a tool path from the tool tip viewpoint. The second type is from a fixed viewpoint but shows the weld tool moving over the workpiece.

## a. Hand-Eye Viewpoints

The most useful view is from a camera mounted on the tool fixture. This provides a close-up view of the tool, joint and nearby fixtures. As this camera is subjected to the motions of the tool frame, the view appears to be a static tool with a moving background, and can be difficult to interpret without practice. This is important because it is the view that the operator is most likely to use.

Precise knowledge of the manipulator's location is needed to determine the extrinsic parameters of the camera. The camera is rigidly fixed to the tool frame and is therefore rolling, pitching and yawing as a result of the motion of all the manipulator joints. The technique used is to establish the camera transformation in two stages:

- Transform global coordinates to tool coordinates.
- Determine the tool frame to camera frame transform (the camera is fixed relative to the tool).

The transform from global to tool coordinates is convenient because the forward transformation (tool to global) is known from modelling of robot kinematics. We need to find the inverse of the forward transformation.

$$T_t^0 = \begin{pmatrix} R_0^t & D_0^t \\ 0 \quad 0 \quad 0 & 1 \end{pmatrix}^{-1} = \begin{pmatrix} (R_0^t)^T & -(R_0^t)^T D_0^t \\ 0 \quad 0 \quad 0 & 1 \end{pmatrix} \tag{14}$$

A global coordinate $p_0$ is transformed into the camera coordinate frame using the following equation

$$p_c = T_c^t T_t^0 p_0 \tag{15}$$

Figure 12 illustrates the modelling environment with two modelled views of a vee-groove workpiece. The first view is static. The second view is from a camera mounted on the end-effector and subjected to roll.

The camera view models were tested in two stages. The first stage was a simple laboratory test to verify the methodology and implementation of algorithms. The second stage of testing was carried out before commissioning vision equipment on the welding manipulator.

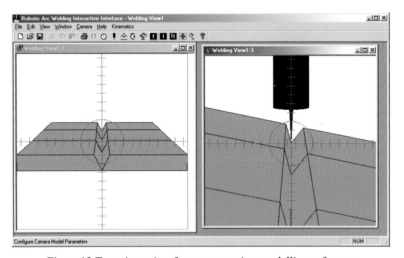

Figure 12 Two viewpoints from camera view modelling software.

## b. Laboratory Test

A laboratory scene featuring a doorway (Figure 13) was chosen, as this was structured (orthogonal and permanent), and the door presented a simple case being planar, with easily measured dimensions. Images were captured using a CCD camera at various orientations and positions of the camera. The images contained partial or occluded views of the doorway.

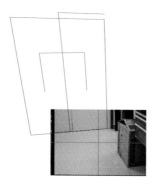

Figure 13 Laboratory test image.

Six extrinsic parameters are required for a camera model. These are:
- The camera displacement from world coordinate origin, $Xc$, $Yc$ and $Zc$.
- The camera orientation, $\alpha$, $\beta$ and $\gamma$.

Intrinsic parameters are needed to model how the image is formed on the CCD, and then how this is represented in computer memory [31]:

- Effective focal length, $f$.
- Dimensions of CCD.
- Number of pixels per mm in $x$.
- Number of pixels per mm in $y$.
- Centre $x$ of image.
- Centre $y$ of image.

Intrinsic parameters were based on manufacturer's data, apart from effective focal length that was determined by calibration. Images were loaded and displayed using Matlab Image Processing Toolbox. A user-defined function allowed manual setting of the camera extrinsic parameters, composition of the transformation matrix and overlay drawing of the door on the camera image data. An Example of this can be seen in Figure 13. The demonstration verified that the models were sufficiently accurate for the proposed application, even though the camera had been calibrated with a very crude manual technique.

### 6.2.4    *Integration of Models and Welding Workpiece Images*

The preliminary results demonstrated the potential of the models and ease of integration with real hardware. In the next stage, cameras were integrated with the 5-axis welding robot. Three camera positions were set up:

- Side angle12 mm lens (normal perspective).
- End angle 25 mm lens (narrow angle).
- Hand-Eye (tool mounted 7 mm lens wide angle close-up).

Images were captured using a Matrox Orion frame grabber. Software was compiled using MIL (Matrox Imaging Library) with Microsoft Visual C++. Figure 14 shows the overlay of geometric model on a static view of the workpiece.

Figure 15 is an overlay of the welding tool model on an image from the camera mounted on the end-effector. The wokpiece model could not be integrated with this view until a link between the robot motion control and image processor has been implemented. The motion control is needed to supply the positional feedback data to calculate the camera's extrinsic parameters.

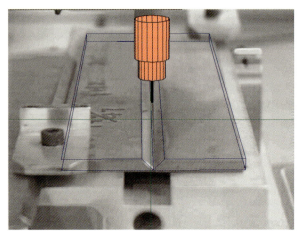

Figure 14 Overlay of tool model on fixed camera image.

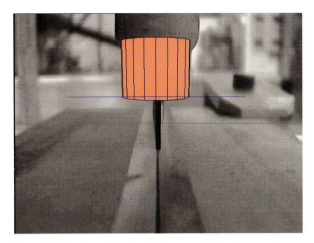

Figure 15 Overlay of tool model on eye-in-hand camera image.

# 7.    Conclusions

The subsystems of a robotic GTAW system for high performance materials such as titanium exist. There remains a reluctance for end-users to accept fully automated systems on account of component cost, system integration cost, and high levels of risk associated with unproven technology.

Increased levels of automation can be achieved by retaining a system supervisor. In this case, the subsystems are acting as tools to assist the operator in decision-making and reducing the dependency on manual skills. This approach will allow a phased implementation of technology. With time, and if product consistency and quality improves, users' confidence will grow and the supervisor will be required to do less and less. Eventually a fully automated system will be realised with a supervisor only required to resolve conflicts or recover from fault conditions. Manual skills will still be required for preparation, fixturing and tack welding.

System integration costs must be reduced. Adoption of standards for exchange of welding data and equipment interfacing will help to achieve this. An open architecture controller will reduce system integration time and conflicts, allowing greater flexibility in choice of subsystems.

The welding demonstrator, commissioned at Gintic, is acting as a test bed for subsystems and their integration in an intelligent system. At present, the features are: the GiCNC open-architecture controller, seam tracking, weld pool monitoring, vision systems and interfaces for process control. The features of a system for titanium welding have been presented.

Kinematics were analysed because a consistent description of the manipulator behaviour is needed by many of the subsystems such as trajectory planning and sensor processors. Process control is a complex problem in arc welding but any system will depend upon a reliable process controller that can cope with the many variables and their interrelationships. These, in turn, depend upon reliable sensory data.

The potential of camera systems in providing operator feedback and integration with geometric modelling has been demonstrated. Their reliability depends upon good calibration and low sensitivity to errors in manipulator positioning.

Further work is planned in the following areas:

- Improving robustness of seam-tracking algorithms to surface reflections.
- Use of seam trackers on more complex weld profiles.
- Auto calibration of visual sensors to reduce dependency on lengthy set-up times and manual calibration.
- Image processing to compare modelled features with detected features to check robot workpiece position anomalies.
- Fusion of camera images with seam-tracker data.
- Neueofuzzy process control.

## References

1.  Chen, X.Q., and Lucas, J., "Sensors for arrow gap welding". Proceedings of Welding Production Procedures Being Technology – Arc Welding 1989, DVS Berichte 1989, No. 127, pp. 101-105.

2.  Chen, X.Q., and Lucas, J. "A fast vision system for control of narrow gap TIG welding". International Conference on Advances in Joining and Cutting Processes, Harrogate, UK, 31 October – 2 November 1989, pp. 8-1 to 8-9.

3.  Shacklock, A., Luo, H., Huang, S., and Wang, J.Y. "Intelligent robotic GTAW system for 3D welding", Gintic Technical Report AT/01/013/AMP, 2001.

4.  "Welding Titanium, A Designers and Users Handbook", TIG The Titanium Information Group, The Welding Institute, 1999.

5.  "Titanium, Design and Fabrication Handbook for Industrial Applications, Timet", Titanium Metals Corporation, 1997.

6. Craig, J.J., "Introduction to Robotics, Mechanisms and Control", 2nd edition, Addison-Wesley, 1989.

7. Wolovich W.A. "Robotics: Basic Analysis and Design", Holt, Rinehart and Winston, 1987.

8. Cary, H. "Modern Welding Technology", 4th edition, Prentice Hall, 1998.

9. Hoffman, T. "Real-time imaging for process control", Advanced Materials and Processes, 140(3), 1991, pp. 37-43.

10. Kovacevic, R. and Zhang, Y.M., "Machine vision recognition of weld pool in gas tungsten arc welding", Proceedings of the Institution of Mechanical Engineers, Part B - Journal of Engineering Manufacture, Vol. 209 (B2), 1995, pp. 141-152.

11. Kovacevic, R., Zhang, Y.M., and Li, L. "Monitoring of weld joint penetration based on weld pool geometrical appearance", Welding Journal 75(10): pp. 317-s to 329-s.

12. Luo, H., Devanathan, R., Wang, J.Y., and Chen, X.Q. "Vision based weld pool geometry control using neurofuzzy logic", Proceedings of Asia Conference on Robotics and its Applications, Singapore, June 2001, pp. 141-146.

13. Richardson, R.W., and Gutow, D.A. "Coaxial arc weld pool viewing for process monitoring and control", Welding Journal 63(3), pp. 43-50.

14. Pietrzak, K.A. and Packer, S. M. "Coaxial vision-based weld pool width control", PED- Vol.51, Welding and Joining Processes ASME 1991, pp. 251-264.

15. Brzakovic, D., and Khani, D.T. "Weld pool edge detection for automated control of welding", IEEE Transactions on Robotics and Automation, Vol.7(3), pp. 397-403.

16. Anderson, P.C. "A review of sensor systems for the top-face control of weld penetration state in GTAW", The International Journal of Machine Tools and Manufacture, Vol.34, No.8, 1994, pp. 1079-1090.

17. Beardsley, H.E., Zhang, Y.M., and Kovacevic, R. "Infrared sensing of full penetration state in GTAW", International Journal of Machine Tools and Manufacture, Vol. 34, No. 8, 1994, pp. 1079-1090.

18. Duley, W.W., and Mao, Y.L. "Effect of surface condition on acoustic emission during welding of aluminium with $CO_2$ laser radiation", Journal of Applied Physics D: Applied Physics 27(7), pp. 1379-1383.

19. Irving, R. "Sensors and controls continue to close the loop in arc welding", Welding Journal, April 1999, pp. 31-36.

20. Quinn, T.P., Smith, C., McCowan, C.N., Blachowiak, E., and Madigan, R.B. "Arc sensing for defects in constant-voltage gas metal arc welding", Welding Journal Research Supplement, September 1999, pp. 322s-328s.

21. Okamura, K. "Ultra High-Speed arc Welding (4m/min)", Industrial Robot, Vol.25, No.3, 1998, pp. 185-192.

22. Pritschow, G., Horber, H., and Haug, K. "Intelligent laser-stripe sensor system for fully-mechanised adaptive multipass arc welding", Proceedings Asia Conference on Robotics and its Applications, Singapore, June 2001, pp. 135-140.

23. Chen, S.B., Lou, Y.J., Wu, L., and Zhao, D.B. "Intelligent methodology for sensing, modelling and control of pulsed GTAW: Part 1 – Bead-on-plate welding", Welding Journal Research Supplement, June 2000, pp. 151s-164s.

24. Chen S. B., Lou Y. J., Wu L. and Zhao D. B., "Intelligent methodology for sensing, modelling and control of pulsed GTAW: Part 2 - Butt Joint Welding", Welding Journal Research Supplement, June 2000, pp. 165s-174s.

25. Zhang, Y.M., and Kovacevic, R. "Neurofuzzy model based control of weld fusion zone geometry", IEEE Transactions on Fuzzy Systems, Vol.6, No.3, 1998, pp. 389-401.

26. Durrant-Whyte, H.F. "Integration, Coordination and Control of Multi-Sensor Robot Systems", Kluwer, 1987.

27. Rippey, W.G. "Industry needs in welding research and standards development", Proceedings of NIST Workshop, April 1996, NISTIR 5822.

28. Rippey, W.G., and Falco, J.A. "The NIST automated arc welding testbed", Proceedings of 7th International Conference on Computer Technology in Welding, San Francisco, CA July 1997.

29. "Manual of Photogrammetry", 4th edition, American Society of Photogrammetry, 1980.

30. Wolf, P.R., and Dewitt, B.A., "Elements of Photogrammetry with Applications in GIS", 3rd edition, McGraw Hill, 2000.

31. Tsai, R.Y. "A versatile camera calibration technique for high-accuracy 3D machine vision metrology using off-the-shelf TV cameras and lenses", IEEE Journal of Robotics and Automation, Vol.A-3, No.4, August 1987, pp. 323-344.

# CHAPTER 8

# LASER MATERIAL PROCESSING AND ITS QUALITY MONITORING AND CONTROL

Hong Luo*, Hao Zeng*, Lunji Hu**, Zude Zhou**,
Xiyuan Hu**, Youping Chen**

*Gintic Institute of Manufacturing Technology,
71 Nanyang Drive, Singapore 638075

**Huazhong University of Science and Technology (HUST), P. R. China

## 1. Introduction

Laser applications have a history stretching almost forty years. Since then both the available laser equipment and laser applications have undergone extensive advancements.

### 1.1 *Laser Equipment*

In material processing, the two most commonly used laser systems are the $CO_2$ gas laser and the Nd:YAG solid state laser. The former includes diffusion-cooled slab laser, fast axial flow laser (DC excited or RF excited) and cross flow laser. For the Nd:YAG laser, which can be equipped with fibre optic beam delivery system, mainly lamp-pumped and diode-pumped lasers are used. Table 1 compares the $CO_2$ and Nd:YAG lasers [1].

Initially developed for CD players, laser printers etc., laser diodes are now being used for high power applications. By using special cooling and stacking techniques, High Power Diode Lasers (HPDLs) can be manufactured. They have already achieved output powers that can compete with conventional gas or solid state lasers. They have a system efficiency of more than 30%, and are extremely compact and maintenance free. The lifetimes of laser diodes are now more than 10,000 hours. Even though they cannot achieve the focusability of conventional lasers, HPDLs will be a suitable laser source for a wide range of applications, such as soldering, welding, and surface treatment. At present the powers range from 10 watts up to 6,000 watts at wavelengths between 790 nm to 980 nm.

Table 1 Comparison of $CO_2$ laser and Nd:YAG laser.

| | $CO_2$ | Nd:YAG |
|---|---|---|
| Power, kW | 15-25 (40) | 2-3 (5) |
| Beam quality up to 500W | Comparable | Comparable |
| Beam quality at high power | Good | Low |
| Power dependent variation of beam quality | Low | Strong |
| Efficiency, % | 10-20 | ≤5 |
| Beam handling | Rigid mirror systems | Fiber |
| Pulse mode | Limited | High energy short pulses |
| Wavelength | 10.6μm | 1.06μm [1] |
| Lifetime | High | ≈400h (lamp) |
| Running cost | Comparable | Comparable |
| Cost of investment | | More expensive |
| Cost of beam handling | | Cheaper |

NB (1) Higher absorption in metals

Today's high power diode lasers achieve spots with dimensions of approximately 1 mm$^2$, delivering intensities breaching $10^5$ W/cm$^2$ and power levels of up to 6 kW. In different applications, different focus sizes to maximise the benefits of the laser source are required [2]. Variable cylindrical optics are designed and manufactured for this purpose. They create rectangular spot sizes with adjustable aspect ratios. These optics are used in connection with high power diode laser stacks operating in the multiple kilowatt power range. Two different variable optics are used. One is designed for surface applications such as hardening, cladding and alloying. The spot size varies from 6.0 mm to 22.0 mm in one direction and stays constant at 2.6 mm in the other direction. The other one exhibits two ranges of continuously variable spot sizes from 1.2 mm to 6.0 mm and 2.4 mm to 12.0 mm with its second dimension fixed at 1.2 mm or 2.4 mm respectively. These lenses are used for applications from surface treatment to welding and cutting.

Examples of successful industrial implementation of HPDLs include plastic welding, surface hardening and heat conduction welding of stainless steel and aluminium [3]. The joining of plastics with an HPDL offers the advantages of producing a weld seam with high strength, high consistency and superior appearance. Another application is the keyless entry system introduced to the Mercedes E-class where the microelectronic circuits are embedded in a plastic housing. Other applications include instrument panels, cell phones, headlights and taillights. Applications in the field of surface treatment of metals profit from the HPDL's inherent line-shaped focus and the homogeneous intensity distribution across the focus. An HPDL system is used to harden rails for coordinate measurement machines. This system contains a customised zoom optics to focus the laser light onto the rails and, with the addition of a temperature control, even complex shapes can be hardened with a constant depth and minimum distortion.

## 1.2    *Applications of Laser Material Processing*

Applications of laser material processing cover machining, welding, cutting, heat-treatment, cladding, re-melting, marking, scribing, drilling, surface-treatment, deposition, forming, ablation, rapid prototyping, etc. The materials that can be processed by laser cover not only metals, such as, steel, copper, aluminium, titanium, and nickel-based superalloys; but also non-metals, such as, inorganic materials (alumina, quartz and glass), organic materials (wood, cloth, plastic), and composites.

An important application is laser blank welding in automotive industries. The laser welded blanks [4], made using this new technology, have been applied to actual body components with satisfactory results in material cost savings, improved body accuracy, and decreased body weight.

The typical car body is made up of more than 300 pieces, with different thickness and treatments, according to the characteristics of the part to be welded. The cost of a car body comprises the material cost 48%; the welding cost 12%; the stamping cost 13%, and the ceiling cost of the joining point 3% [4]. Thus it is very important to increase the material yield and to decrease the spot welding, the stamping dies and the joining point for car body cost reduction.

If these separate panels are stamped to form a one-piece panel efficiently, by using integrated sheet blanks, we can reduce the costs of the material, welding, stamping and ceiling. "Laser welded blank" which joins different kinds of sheets before stamping (Blanking—Welding—Stamping) has overcome the disadvantages of two conventional blank methods, namely divided type and one-sheet type.

When welded sheet metals are used for stamping, the welding characteristics must satisfy the requirements of joint shape, high strength and high productivity. $CO_2$ laser welding can meet all these requirements. Furthermore, the $CO_2$ laser welds have a smaller hardened area and no heat-softened area. This is very important for press forming because the hardened area decreases the formability and the softened area permits too much extension and may cause a partial crack.

A laser welding line, consisting of 5 stations and 2 or 4 laser oscillators, has been established. The line can produce different panels by changing the jigs. It has been used for welding large panels, for example, the side member.

With the rapid development of laser equipment and due to the advantages of the laser techniques, laser material processing is finding more and more applications in various industries.

## 1.3    *Automation of Laser Material Processing*

Some significant advantages of laser material processing have powered it towards automation.

Firstly, it is very flexible in the way it can be programmed to direct the laser energy via a robotic beam delivery system.

Secondly, there is very little environmental disturbance in delivering laser energy. For example, there is no electric field, no magnetic field, no heat, and no mechanical stress. Thus any signal in these areas will probably have to come from the process itself. This gives a wide window for in-process diagnostics unique to the laser.

In a closed-loop control system (Figure 1), the control sequence is as follows:

1.  A process variable or product quality is measured.
2.  The signal is compared to the desired value and an error detected.
3.  This error initiates a change in the process manipulators or drives, thus affecting the process.
4.  What is changed, and by how much, is decided by the controller.

The controller is able to take the signal from the process while the process is running, and compute sufficiently fast so that the error detection can be made, and the machine corrected before there is any considerable product waste. In-process sensing and signal processing is fast becoming one of the strengths of the laser material processing.

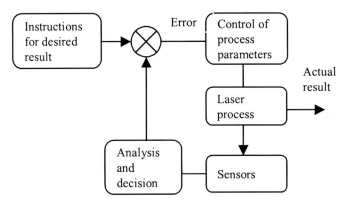

Figure 1 Diagram of the structure of a closed-loop control circuit.

### 1.3.1    In-Process Monitoring

For the control or monitoring of laser material processing, the in-process signals listed in Table 2 are required [5]. Table 3 lists some of the main concepts being investigated [5].

Table 2 Principal variables that characterise a laser process.

| Beam | Workstation | Workpiece |
|---|---|---|
| Power | Traverse speed | Surface absorptivity |
| Diameter | Vibration, stability | Seam location |
| Mode structure | Focal position | Temperature |
| Location | Shroud velocity and direction | Quality of product |

### a.    Monitoring Beam Characteristics

A number of techniques have been patented or developed for the in-process monitoring of a laser beam, including laser beam analyser (LBA), perforated mirror, chopper devices, heating mirror, heating wire, photon drag in germanium (Ge), piezoelectric, heating gas, optical scattering, and acoustic signals.

Table 3 Some in-process sensors currently under investigation or in the market.

| Signal | Sensor | Availability |
|---|---|---|
| Beam power | Laser beam analyser (LBA), Leakage from cavity mirror | Commercial |
| Beam diameter and mode | LBA, Hollow needle, Perforated mirror | Commercial Commercial |
| Location | Acoustic mirror, LBA, Edge thermocouples, Modified LBA, Scanning beam splitter | Research Commercial Research Research |
| Vibration or stability | Accelerometers, Strain gauges Laser doppler anemometer (LDA) | Commercial Commercial |
| Focal position | Infrared, Pressure Capacitance, Inductance | Research Commercial |
| Shroud gas velocity | Nozzle pressure Schlieren, Speckle interferometer | Commercial Research |
| Surface absorption | Acoustic mirror Back reflection | Research Research |
| Seam location | Optical Pressure, Acoustic | Res./Comm. Research |
| Cutting quality | TV camera on spark discharge Temperature of cut face Acoustic mirror Viewing down beam | Research Research Research Research |
| Welding quality | Acoustic mirror Acoustic workpiece Sonic microphone Optical emissions Electric signals Plasma charge sensor Laser probe Acoustic nozzle Video camera | Research Research Research Research Research Research Research Research Research |
| Surface hardening quality | Temperature Infrared, Acoustic | Res./Comm. Research |
| Cladding dilution | Inductance | Research |
| Powder feed rate | Pressure, Stress, Vibration | Research |

## b.    Monitoring Work Table Characteristics

The worktable variables are fairly straightforward as in the measurement and control of position, traverse speed or nozzle gas velocity. Traverse speed and table position can be monitored by encoders, tachometers, laser doppler anemometer (LDA), and laser interferometer. Vibration and stability can be monitored by accelerometers and strain gauges. Shroud gas velocity can be monitored by nozzle pressure, Schlieren and speckle interferometry.

The focal position is crucial, and subtler to measure. It needs to be measured in real time because the workpiece may warp slightly during processing or the part may be contoured. The programming of the line to be followed may be simplified if the height above the workpiece is controlled in process rather than left to precise and time consuming pre-programming. There are several signals that are used currently for sensing and controlling the height of the nozzle above the workpiece, such as infrared, capacitance, inductance and pressure.

In laser welding, with the narrow fusion zone associated with laser welds, there is a need for a seam following system which is accurate and fast. Seam tracking systems with diode lasers and CCD cameras are used to locate the seam point.

## c.    Monitoring Workpiece Characteristics

For laser cutting, TV cameras on spark discharge, temperature of cut face, acoustic mirror and viewing down beam have been adopted for quality monitoring.

For surface hardening quality, temperature and infrared are used, while for cladding dilution, inductance is investigated.

For laser welding, various techniques have been explored, such as: acoustic mirror, sonic microphone, optical emissions, electric signals, plasma charge sensor, laser probe, acoustic nozzle and video camera. All of these will be discussed in detail in the following section.

### 1.3.2    In-Process Control

The first step is to obtain the diagnostic signals, and the next is to use them in a closed-loop control system. The control system comes in two basic schemes: single-input-single-output (SISO) control and multiple-input-multiple-output (MIMO) control. In a SISO control scheme, there is one cause and one effect. For example, if the power is too low, raise the current to the discharge tubes. If it is too high, do the opposite. In an MIMO control scheme, there are multiple diagnostic signals and more than one interrelated operating conditions to be adjusted. This type of control system requires decision-making software which could constitute intelligent processing. The main components of such a control system are illustrated in a smart laser cutting system (Figure 2) [5].

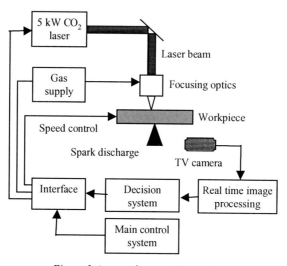

Figure 2 A smart laser cutting system.

The decision-making can be performed by fuzzy logic whereby a probability matrix determining which parameter is the most likely cause of the fault condition has its values adjusted by the program itself. This is a form of self-teaching. An alternative technique is to control the process using a neutral network.

## 2.    Survey of Real-Time Laser Welding Quality Monitoring

In recent years, laser welding has found increasing applications in various industries, especially in automotive production. It has several advantages when compared to other welding processes such as arc welding. These advantages include deep penetration, high depth to width ratio, and low distortion. A primary concern in the industry is the detection of weld defects using real-time monitoring methods, which must be reliable, flexible, and cost-effective in high capacity non-clean environments.

Laser keyhole welding is a very complex process influenced by a lot of factors and interactions, such as: the melting and vaporisation of materials, keyhole formation and shape, laser-plasma-materials interaction and energy absorption through inverse Bremsstrahlung in the plasma and Fresnel absorption at the keyhole wall. Fortunately these phenomena also provide many sources of information for monitoring the welding process itself.

Typical sensors used to monitor laser welding include acoustic emission, audible sound, weld pool infrared emission, voltage difference between workpiece and nozzle, visible light and ultraviolet (UV) emissions from laser-induced plasma. Two sensors that have not been extensively used for laser weld monitoring but show some promise are electromagnetic acoustic transducers and polyvinylidene flouride detectors. Figure 3 summarises various sensing methods used for real-time laser welding monitoring.

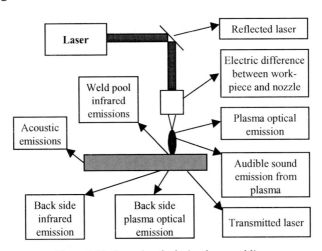

Figure 3 Various signals during laser welding.

## 2.1  *Acoustic Emission*

Acoustic emission (AE) monitors stress waves induced by changes in the internal structure of a part. As the part undergoes material changes during and after the welding process, AE signals are generated which can be related to the weld characteristics. AE is well suited for real-time monitoring of weld structure changes such as martensite formation; and weld defects, such as, lack of fusion, and centre-line crack.

Adequate weld penetration is critical for good welds. To ensure adequate penetration, it is essential to maintain a keyhole. When the laser welding is conducted under keyhole mode, the laser reflection from the workpiece to the optics mirror is very low. Otherwise when laser welding is carried out under conduction mode, the laser reflection is very high. Therefore, not only welded workpiece and nozzle, but also the output mirror in the laser can generate AE signals. All these signals can be used to monitor the laser welding process [6-10].

Piezoelectric or capacitive transducers are typically employed to sense AE stress waves by converting the resulting displacements on the surfaces of the structure into electrical signals. These sensors are usually put on the workpiece, nozzle and optical mirrors in the output window of the laser.

One approach was to measure the thermal stress formed during laser spot welding using a quartz crystal transducer coupled to the workpiece [9-10]. The results showed that a strong AE signal was produced during keyhole welding.

Another approach was to monitor the AE signals from the output mirror in the laser [6-8]. Steen and Weerasinghe showed that keyhole formation could be monitored using an acoustic mirror to measure AE signals on the mirror. The results showed that there was a rise in the acoustic signal as the keyhole began to fail.

A disadvantage of using AE monitoring is the need for the transducer to be in contact with the workpiece or mirror. This could cause difficulties due to the hostile environment of the weld zone, and possibly, an inadequate coupling between the transducer and the workpiece. Even though acoustic emission continues to interest academy and industry, research activity in the field has declined. As a result, non-contact audible sound sensors have drawn increased attention in recent years.

## 2.2    *Audible Sound*

Audible sound, also referred to as airborne acoustic emission, encompasses the human audible range of 20 Hz ~ 20 kHz. Sources of sound emission are normally from the weld zone (thermal stresses, pressure waves caused by plasma and metal vapour, and cracking) but can also be influenced by external sources (gas jets and ambient noise). These sounds can be correlated to weld states such as weld penetration. Audible sound detection is non-contact, thus relieving some of the difficulties associated with contact detection.

A typical sensor for audible sound is a condenser microphone positioned in the vicinity of the weld zone.

Major results using audible sound to monitor welding quality are listed below:

- High airborne emission activity was detected for a good weld while low activity was detected for a bad weld [11].
- Audible sound signal strength was a function of penetration depth. This indicated that the audible sound signal amplitude increased dramatically when a keyhole was formed [12].
- Typical failures such as porosity, poor penetration, and butt joint mismatches were detectable but the system could not distinguish between different failure types [13].
- Frequency analysis of audible sound emissions by Fast Fourier Transformation was further explored. Different results were achieved by different researchers under different conditions.

Gu and Duley [14-16] used audible sound emission near 4.5 kHz to identify weld conditions. Three classes were investigated using a linear discriminant algorithm for classification. Results were 83% correct identifications for overheated, 67% for full penetration, and 75% for partial penetration.

For Farson et al [17], the total energy in the bandwidth of 1 kHz ~ 2 kHz was compared with a suitable threshold value. Welds falling above the threshold were classified as full penetration and those below the threshold, partial. Results showed 90% classification for insufficient penetration and near 100% for complete penetration.

Audible sound detection has become one of the most promising techniques recently investigated for on-line monitoring of laser welding. Another promising method is infrared sensing.

## 2.3    *Infrared Sensing*

All objects above absolute zero temperature, whether solid, liquid or gas, emit electromagnetic energy, with hot objects radiating more electromagnetic power than cool objects. Infrared sensing encompasses about 0.72~1000 μm of the electromagnetic spectrum. By relating infrared radiation from the weld zone to weld states, weld quality can be determined.

In laser welding, infrared detectors provide information on the temperature of the weld pool and surrounding metal. Monitoring the temperature enables several features of the weld to be monitored, for example, bead width and penetration. Several set-ups have been applied including direct viewing, coaxial viewing, and backside viewing.

CCD cameras could be used to obtain the thermal images of laser weld pool. However, due to the interference from the bright laser-induced plasma, the images are not always good enough and it takes too long to process such images.

Recent investigations have focused on infrared signal analysis (temperature measuring), where infrared signals are converted to electrical signals and analysed using signal processing techniques. Nava-Rudiger and Houlot [18] used two infrared photodiodes to detect geometric defects during laser welding of steel. They concluded that weld defects, such as gap and misalignment, could be detected by adding and subtracting the signals,

Lankalapalli et al [19] proposed a model to correlate weld penetration depth to the weld width on the top surface and the temperature at the bottom surface. The model is computationally efficient and estimates penetration depth from easily measurable weld bead width and surface temperature.

Infrared sensors can monitor thermal signals but are unable to sense plasma fluctuations which primarily occupy the visible and ultraviolet spectrum.

## 2.4    *Ultraviolet Sensing*

Radiation associated with the plasma is concentrated in the ultraviolet (UV) and visible band of the electromagnetic spectrum. Laser-induced plasma plays a significant role in understanding the keyhole dynamics during laser welding. However, the level of comprehension is still quite limited. Plasma temperature strongly influences the electron density in the plasma. The plasma temperature, however, is maintained by absorption of the incident energy by the free electrons in the plasma. This results in a significant portion of ultraviolet radiation being emitted by high temperature plasma. Thus, changes in the temperature of the plasma emitted from the keyhole will result in a change in the UV signal, which can be used to detect the state of the weld.

UV sensing is usually correlated with keyhole formation and weld penetration. Set-ups for UV monitoring are similar to IR set-ups. There are two major applications of UV sensing: (1) spectroscopic analysis and (2) UV signal analysis.

## 2.4.1   Ultraviolet Spectroscopic Analysis

Spectroscopic analysis is useful for determining the ion content in the plasma. This is important in determining the extent of plasma formation due to metal vapour and shielding gas. Spectroscopic studies in laser welding are concentrated in the investigation of plasma physics and plasma-material interaction.

Rockstroh and Mazumder [20] studied plasma-material interaction during continuous laser welding. Spectroscopic measurement was used to obtain two-dimensional mapping of plasma parameters for calculating absorption, refraction, and radiation losses. Results showed that 56%~80% of the non-plasma flux was delivered to the target for laser powers between 5 and 7 kW. Larger spot sizes and heat-affected zone (HAZ) were attributed to refraction of the laser through the plasma that enhanced coupling and resulted in deeper penetration.

Spectroscopic methods were also used to determine the plasma temperature. Results in welding steel showed that the plasma temperature corresponds to weld penetration depth. Monitoring the laser weld plume spectrum of mild and stainless steels, Muller [21] used a computer controlled optical spectrometer with fibre optic input to determine the temperature and concentration of major elements in the plume. The results showed that the UV signal depended on laser power; the higher the power, the higher the temperature of the plasma. Furthermore, slower welding speeds exhibited larger short-term fluctuations in temperature. No change in iron ion concentration was detected for different speeds.

Even though spectroscopic analysis is very important for understanding the behaviour of plasma, it is not suitable for on-line quality monitoring because this method is costly and time consuming. It is mainly used in theoretical analysis.

## 2.4.2   Ultraviolet Signal Analysis

While spectroscopic analysis is still very important, recent developments have shown much potential in ultraviolet signal analysis. In ultraviolet signal analysis, ultraviolet signals from the weld zone are converted into electric signals. Magnitude and frequency analyses of the input signals are then correlated with weld states.

Plasma formation and weld penetration are closely related. When penetration occurs, a clear relationship between full penetration and the ultraviolet plasma signal frequency band ratio is observed. Taking advantage of this observation, Beyer et al [22] monitored plasma fluctuations using a streak camera and the outputs were correlated with weld penetration. A neural network algorithm processed the camera output and process parameters (e.g. beam power, focal position, etc.). The results showed that full penetration could be detected 98.5% of the time, and detection of gap size greater than 0.2 mm was 99%.

Kluft et al [23] used ultraviolet sensors to monitor plasma radiation from the weld by directing an optical cable towards the plasma radiation. It was indicated that the main advantage of the set-up was that spray, vapours, detector contamination, and detector sediment did not change the signal ratios, making the system suitable for hostile environments. Besides weld penetration, surface defects could be monitored by observing the plasma fluctuations in the weld zone using a CCD camera. It was concluded that a decrease in light intensity emitted from the plasma core indicated underfill or pits in the weld.

Developments in ultraviolet sensing were incorporated in industrial applications by Kaierle et al [24] who introduced an autonomous manufacturing unit for three-dimensional (3D) laser welding. Diagnostic sensors such as measurement of laser power, beam position, power density, and beam profile, were coupled with temperature and plasma detection. The system was first trained with good and bad weld states and then run on line.

Non-contact sensors are attractive for weld monitoring due to the hostile nature of the weld zone. Remote sensing using ultraviolet detectors accomplishes this task while maintaining a safe distance.

Other sensors include the plasma charge sensor (PCS). Its signal comes from an electric space charge associated with the plasma, identified by Li et al [25-26]. The signal can be used to diagnose the general health of a laser weld. Together with an acoustic sensor, the plasma charge sensor has successfully been used to measure and log welding faults on a laser can welding unit.

In addition, EMAT detectors and polyvinylidene fluoride are a new form of detectors that have potential for laser process monitoring [27].

By combining different kinds of sensors into a multi-sensor system, the advantages of each sensor are incorporated into one system. Two types of sensor fusion techniques have been extensively investigated in laser welding monitoring. These are infrared [28-30] and ultraviolet detection, sound and ultraviolet detection [31-33].

From the years of experience of the authors in laser welding, it has been observed that when bright blue plasma and some audible sound occur steadily during laser welding, the process must be done under keyhole mode and good quality can be achieved.

Efforts had been put in the analysis of the optical and audible acoustic signals emitted from the laser-induced plasma and these led to a monitoring method combining visible light sensing and audible sound detection.

## 3.     Analysis of Optical and Acoustic Signals Emitted From Plasma and Sensor Design

It has been shown that keyhole formation and stability play an important role in laser welding. The behaviour of laser-induced plasma is closely related to the formation of keyhole, and thus a good indicator of laser welding quality. The plasma also provides much information that can be used to monitor laser welding quality.

This has been done by various existing monitoring approaches, in which optical and acoustic emissions were used. One problem concerning these methods is that optical and acoustic emissions from the plasma were not analysed thoroughly to obtain the characteristic signals representing the behaviour of the plasma. There is also a lack of theoretical foundation for these monitoring methods. On the other hand, no thorough experimentation concerning various laser welding defects had been carried out, so identifications of such defects are not available yet. The following steps were taken in this study.

Firstly, analyse the optical and acoustic emissions from the laser-induced plasma in order to get the characteristic signals of the plasma.

Secondly, find a theoretical relationship between these signals with the laser welding quality.

Thirdly, experiments, on how these signals change when different defects occur, will be explored extensively using FFT and wavelet analyses.

Finally, a backpropagation (BP) neural network will be trained with the experimental data and used to identify different kinds of defects.

## 3.1   *Laser-Induced Plasma During Welding of Thin Metal Sheets*

All the experiments were carried out under the following conditions:

- Axial fast flow $CO_2$ laser with a maximum power of 1500w
- Welding speed from 2 m/min to 5 m/min
- ZnSe focusing lens with a focal length of 150 mm to focus the incident laser into a small spot with a diameter of about 0.2 mm
- Shielding gas: Argon with a flow rate of about 8 L/min.

The formation of keyhole and laser-induced plasma is a sign of deep penetration welding. In order to prove the existence of plasma during keyhole laser welding, an optical multi-channel analyser (OMA) had been used to analyse the optical emissions during the conduction mode and keyhole mode laser welding. The results are shown in Figure 4.

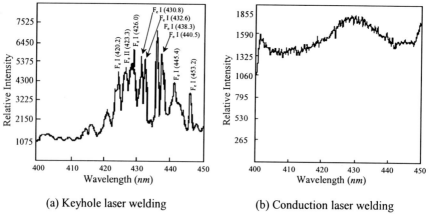

(a) Keyhole laser welding                    (b) Conduction laser welding

Figure 4 Optical emissions during keyhole and conduction laser welding.

In the process of conduction mode laser welding, continuous spectra were observed with little distinction from the background spectrum. In the spectra of keyhole laser welding, transition emissions of excited Fe atoms and Fe I ions could be found. It is evident that metal vapours during keyhole laser welding were partially ionised. According to spectral analysis methods, average electron temperature and electron density in the ionised metal vapours could be computed from the following two equations.

$$T_e = \frac{\hbar(\upsilon_2 - \upsilon_1)}{\kappa * \ln \frac{I_2}{I_1}} \tag{1}$$

where $\hbar$ is Planck constant; $\upsilon$ and $I$ are, respectively, intensity and frequency of the line spectrum and $\kappa$ is Boltzmann constant.

Two line spectra of 445.4 nm and 453.1 nm were chosen to calculate the temperature of electrons. It could be concluded that $T_e = 7006° K$.

$$N_e = 4.83 * 10^{15} T_e^{3/2} * (\frac{g^+ A^+ \upsilon^+ I}{gA \upsilon I^+})$$
$$* \exp(\frac{(E - E^+ - U_1)}{\kappa T_e}) \tag{2}$$

where $T_e$ represents electron temperature; $N_e$ represents electron density; $g^+$, $A^+$, $\upsilon^+$, $I^+$ and $E^+$ represent respectively statistic weight, transition probability, frequency, emission intensity and upper energy level of a first ionised iron ion. $g$, $A$, $\upsilon$, $I$, $E$ represent respectively those of an excited iron atom and $U_1$ represents first ionisation energy of iron atoms.

Iron atoms with a wavelength of 426.0 nm and the first ionised iron ions with a wavelength of 423.3 nm were chosen to calculate electron density. It could be concluded that $N_e \approx 10^{15}$ cm$^{-3}$.

It had been shown that the metal vapour during laser welding was partially ionised. However, not all ionised metal vapour could be identified as plasma. It must meet three criteria below:

1.  If $L$ is the characteristic linear length of the metal vapor, it must be far greater than Debye length $\lambda_D$. $\lambda_D = 6.9(T_e/N_e)^{0.5}$ cm.
2.  Particle density inside Debye Ball $N_D$ must be far greater than 1. $N_D = 1380T_e^{1.5}/N_e^{0.5}$.
3.  If $\omega_p$ represents angular frequency of the fluctuation of the metal vapour, $\tau$ represents average time of collision between atoms and ions, then $\omega_p \tau \gg 1$.

In the present case, $\lambda_D = 18 \times 10^{-6}$ cm. During laser welding, the keyhole penetrated through the thin metal plate, so the linear length of the metal vapour should be in the range of mm. It is obvious that the linear length is much longer than $\lambda_D$.

It was calculated that $N_D \approx 26 >> 1$, so the metal vapour in laser welding satisfies the second condition.

The fluctuation frequency of the metal vapour could be calculated from equation below:

$$\omega_p = {N_e e^2}\Big/{\varepsilon_0 m_e} \tag{3}$$

where $\varepsilon_0$ is permitivity of free space; $m_e$ represents electron mass and $e$ represents the charge of an electron. Hence, $\omega_p \approx 1.79 \times 10^{12}$ Hz.

Since the average time of the collision between atoms and ions is in the range of $10^{-9}$ s, $\omega_p \tau >> 1$. Therefore the metal vapour satisfies the third condition.

The above analysis proves that the partially ionised metal vapour during laser welding of thin metal sheets belongs to plasma. Such plasma is created through ionisation of metal vapour by free electron impact.

## 3.2    *Optical Emission from Laser-Induced Plasma*

There are three kinds of radiation in plasma: free-free transitions, free-bound transitions and bound-bound transitions.

Free-free transitions: continuous spectra occur when free electrons lose part of their kinetic energy in the Coulomb field of positive ions. This loss in kinetic energy is converted into radiation reflecting the Maxwellian distribution of the electrons. The emitted energy is continuous, called "bremsstrahlung," which is typically in the infrared.

Free-bound transitions: because a free electron may assume non-quantised kinetic energies, its recombination with an ion will result in continuous radiation.

Bound-bound transitions: due to their electronic excitation, atoms and ions emit a spectrum of lines such that

$$E_2 - E_1 = \hbar \upsilon_{21} \tag{4}$$

where $E_1$ and $E_2$ are, respectively, the upper and lower excited levels between which the transition takes place; $\hbar$ is Planck constant; $\upsilon_{21}$ represents transition frequency from energy level 2 to energy level 1.

The radiation frequency $\upsilon$ is characteristic of both the atom or ion and the emitting levels. Electrons changing their orbits remain bound to the nucleus, and this type of transition is called a "bound-bound" transition. The corresponding wavelengths (line emissions) extend from the infrared to the far ultraviolet.

These three transitions were all observed in laser-induced plasma from the captured spectra during laser welding. It is known that the strongest line emission from iron particles lies between 400 nm and 440 nm, which was also identified in the captured spectra.

The emission intensity from "bound-bound" transitions is such that

$$I \propto \hbar \upsilon_{21} A_{21} N_2 \Big/ 4\pi \tag{5}$$

where $A_{21}$ represents transition probability for energy level 2 to energy level 1; $N_2$ represents number density of particles in energy level 2.

When energy level 2 and 1 are set, $A_{21}$ and $\upsilon_{21}$ become constants, the intensity of line emission is dependent on $N_2$. $N_2$ can be stated as

$$N_2 \propto N_0 * \exp\left( -(E_2 - E_0) \Big/ \kappa T \right) \tag{6}$$

where $N_0$ represents the number density of particles in ground state.

It is concluded from the above equations that the intensity of line emissions from laser-induced plasma is dependent on the temperature of plasma and the amount of metal vaporisation. The higher the temperature and the more the metal vaporises during laser welding, the stronger the emissions. As such, welding quality could be monitored through the intensity of plasma emission from 400 nm to 440 nm.

## 3.3    *Waves in Plasma*

When laser welding is conducted under keyhole mode and quality welds are achieved, a continuous and stable sound can be heard. When the laser welding process encounters some problems, the audible sound becomes unstable or disappears totally. Its spectra during keyhole mode and conduction mode laser welding were recorded by a sound level meter. The results are shown in Figure 5 and Figure 6, respectively.

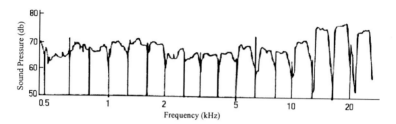

Figure 5 Sound emission during keyhole laser welding.

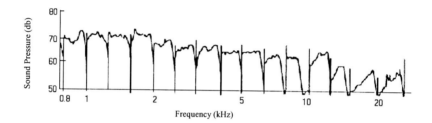

Figure 6 Sound emission during conduction laser welding.

Sound spectra during conduction mode laser welding are almost the same as the background noise. The difference between keyhole and conduction laser welding is in the high frequency range. The sound intensity in the high audible frequency range from 10 kHz to 20 kHz is much higher in keyhole laser welding than that in conduction mode laser welding.

The audible sound wave in plasma comes from two parts: ion sound waves and atom sound waves.

For atom sound waves, the sound velocity could be expressed such that

$$V_0^2 \propto P_0 \Big/ \rho_0 \tag{7}$$

For ion sound wave, the sound velocity ($V_0$) could be stated such that

$$V_i^2 \propto P_e \Big/ \rho_i \tag{8}$$

$$V_0^2 \propto T_0 \Big/ M_0 \quad \text{and} \quad V_i^2 \propto T_e \Big/ M_e \tag{9}$$

so

$$P_0 \propto T_0 N_0 \quad \text{and} \quad P_e \propto T_e N_e \tag{10}$$

where $P_0$ is sound pressure of atom sound; $\rho_0$ mass density of atom; $V_0$ velocity of atom sound; $N_0$ number density of atoms in plasma; $P_e$ sound pressure of ionic sound; $\rho_i$ mass density of ion; $M_i$ mass of ion, $V_i$ velocity of ionic sound; $N_i$ number density of ions in plasma (because particles are only first ionised, so $N_i = N_e$).

During conduction mode laser welding, the sound pressure can be expressed as:

$$P = P_0 \propto T_0 N_0 \tag{11}$$

During keyhole laser welding, the total sound pressure is:

$$P = P_0 + P_e \propto (T_0 N_0 + T_e N_e) \tag{12}$$

As such, the sound pressure in keyhole mode is much higher than that during conduction mode laser welding. The more the metal vaporises, the stronger the sound signal. The total sound pressure is in direct ratio to the number density and temperature of both the atoms and the electrons in the plasma.

Besides, the variations of the sound pressure are also influenced by the velocity of the sound waves. The velocity is defined as the thermal motion velocity of particles in the plasma. Thus the variations of the sound pressure are related to the blasting state and velocity of the plasma coming out of the weld pool and the keyhole. If the shape of the weld pool and keyhole changes, the sound pressure will change correspondingly.

It can be concluded that visible optical emissions corresponding to the strongest line emissions of iron particles (including atoms and ions) and sound pressure in high audible frequency ranges from 10 kHz to 20 kHz can be used to monitor laser welding process and distinguish different kinds of defects.

## 3.4    *Design of Optical and Acoustic Sensors*

For optical emissions, a detector was designed as shown in Figure 7. The chosen diode (3DU80B) is highly sensitive to visible light from 400 nm to 440 nm.

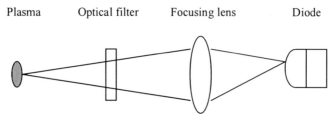

Figure 7 Optical detector.

For audible acoustic signals, an electret microphone (CZII-65), which was direction sensitive (60 degrees), was used to capture high frequency audible sound (10 kHz - 20 kHz) and to eliminate noises coming from other directions.

For optical and acoustic signal processing, the circuits were designed as shown in Figure 8.

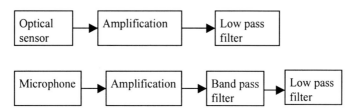

Figure 8 Analog signal processing for optical and acoustic emissions.

## 4.      Signal Processing through FFT and Wavelet Analysis

FFT (Fast Fourier Transform) and Wavelet analysis were used to analyse signal characteristics of different kinds of defects.

The experimental set-up is illustrated in Figure 9. The optical detector and microphone were kept at 45° to the horizontal axis. The distance between the welding pool and the sensors was 200 mm. The data acquisition system had an A/D card (AdvanTech PCL-818HD) with a maximum sampling rate of 100 kHz. The signals were preprocessed before being fed into the data acquisition system.

The experiments aimed to produce full penetration welds as well as defective welds. There were four kinds of defects in these experiments. They are partial penetration due to higher welding speed, too large gap, misalignment, and burn-through by lowering welding speed. The gaps between two sheets were obtained by producing 0.08 mm and 0.12 mm notches at the edge of the sheets. The right and left sheets were staggered slightly in misalignment welds.

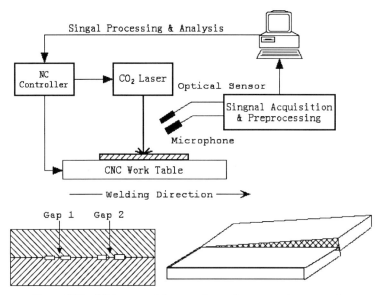

Figure 9 Block Diagram of the experimental set-up and test samples.

## 4.1    *FFT Analysis*

### 4.1.1    *Frequency Characteristics of Optical and Acoustic Signals Using Magnetically Restrained Discharge Laser*

The experiments were conducted under the conditions: magnetically restrained discharge laser; laser power 900 W; welding speed 1m/min; workpiece 0.75 mm thick zinc-coated plates.

The signals are shown in Figure 10 and Figure 11. (Because there was no plasma whose line emissions are from 400 nm to 440 nm, the optical signal during conduction mode laser welding should be zero). The corresponding frequency spectra are shown in Figure 12 and Figure 13 respectively.

It can be seen that the signal intensity increases significantly during keyhole laser welding. Furthermore, there exists a line spectrum of about 300 Hz in the spectra of both optical and acoustic signals.

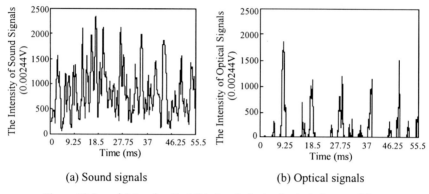

(a) Sound signals                          (b) Optical signals

Figure 10 Sound (a) and optical (b) signals during keyhole laser welding.

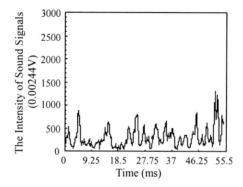

Figure 11 Sound signal during partial penetration.

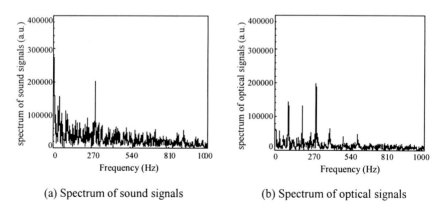

(a) Spectrum of sound signals          (b) Spectrum of optical signals

Figure 12 FFT spectrum of sound and optical signals during keyhole laser welding.

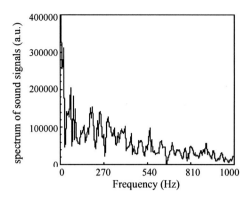

Figure 13 Frequency distribution of sound signal during partial penetration welding.

It should be stated that this kind of phenomena only exists during laser welding using the special laser designed by Huazhong University of Science of Technology (HUST). This is due to the fact that the power source for the laser was not filtered after three-phase rectification. There existed a 300 Hz fluctuation in the output laser power, which led to the fluctuation of laser-induced plasma with the same frequency. The experiments proved that frequency characteristics of both optical and acoustic signals would be more efficient in monitoring of laser welding quality.

### 4.1.2    *Frequency Characteristics of Optical and Acoustic Signals of Differnet Defects*

All the experiments were carried out under the same conditions as stated in Section 3.1. The sound and optical signals and their frequency distributions for keyhole mode laser welding are shown in Figure 14 and Figure 15.

The FFT results show that the main fluctuation of laser-induced plasma is about 400 Hz. However, the main fluctuation frequency is not stable under different experiments even though the same laser welding equipment and the same parameters were used. The instability was caused by the complexity and random nature of the plasma since there are so many parameters that can influence laser welding process in one way or another.

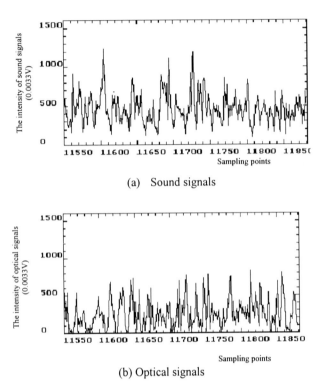

(a)   Sound signals

(b) Optical signals

Figure 14 Sound (a) and optical (b) signals during laser welding at 2m/min.

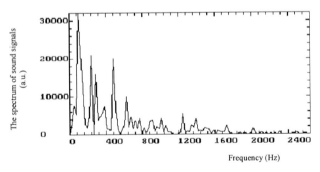

(a) Spectrum of sound signals

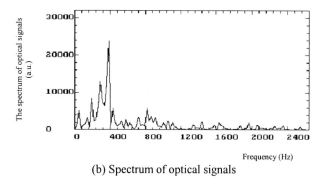

(b) Spectrum of optical signals

Figure 15 Frequency distributions of sound (a) and optical (b) signals at 2m/min.

## a.   *Influence of Welding Speed*

The sound and acoustic signals and their frequency distributions for laser welding at higher speed (4 m/min) are shown in Figure 16 and Figure 17.

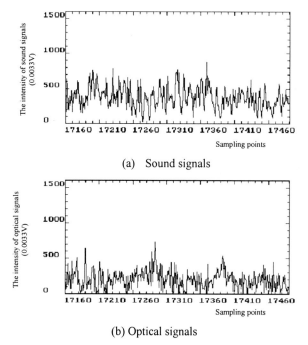

(a)   Sound signals

(b) Optical signals

Figure 16 Sound (a) and optical (b) signals at 4m/min.

As welding velocity increased, since the welded material melted less, the intensity of both signals dropped, but the intensity of the main frequency (from 200 Hz to 1000 Hz) in the frequency domain went down more obviously. Distinguishing of welding defects would be easier if the frequency characteristics of the signals are also used.

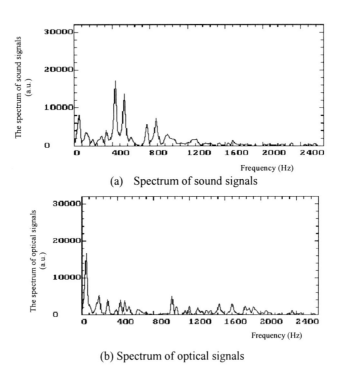

(a)   Spectrum of sound signals

(b) Spectrum of optical signals

Figure 17 Frequency distribution of sound (a) and optical (b) signals at 4m/min.

## b.     *Gap in Butt Joint*

The sound and acoustic signals and their frequency distributions in the presence of a gap in butt joint are shown in Figure 18 and Figure 19.

As the gap increased, since the melted material became less and the shape of the weld pool and keyhole changed, both the time domain and frequency signals changed significantly. The variations of the optical signals were more obvious because some of the plasma went out from the back.

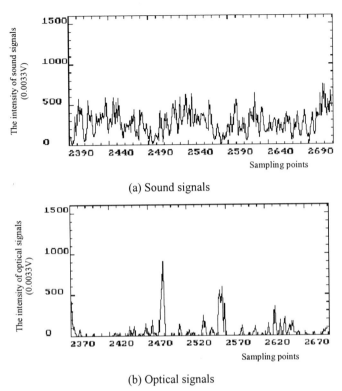

(a) Sound signals

(b) Optical signals

Figure 18 Sound (a) and optical (b) signals with a 0.08 mm gap.

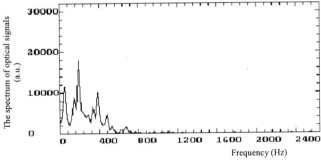

(a) Spectrum of sound signals

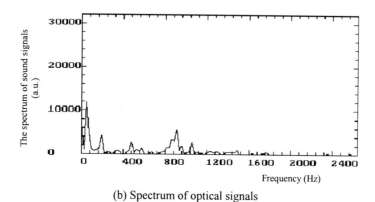

(b) Spectrum of optical signals

Figure 19 Frequency distribution of sound (a) and optical (b) signals with a 0.08 mm gap.

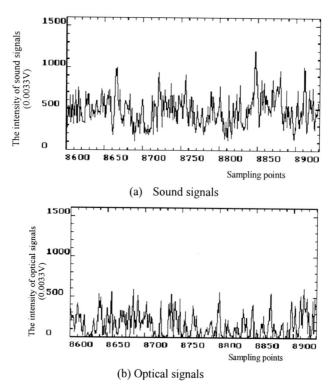

(a)   Sound signals

(b) Optical signals

Figure 20 Sound (a) and optical (b) signals with a misalignment of about 0.2 mm.

## c.    Misalignment in Butt Joint

Both optical and acoustic signals and their frequency distributions in the presence of some misalignments in butt joint are shown in Figure 20 (above) and Figure 21(below).

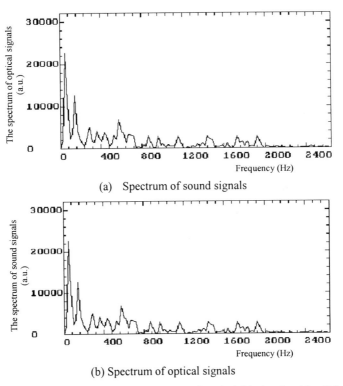

(a)    Spectrum of sound signals

(b) Spectrum of optical signals

Figure 21 Frequency distribution of sound (a) and optical (b) signals with a 0.2 mm misalignment.

If the amount of the misalignment was small, the intensities of both the optical and acoustic signals did not vary obviously because the amount of melted material did not change so much. But the corresponding spectrum distributions changed since the weld pool and keyhole shape changed. This is especially true for acoustic emission because the sound pressure was also influenced by the shape of weld pool and keyhole. All these demonstrated that because of the variations of the weld pool and keyhole, the plasma fluctuation behaviour and its spectrum distributions changed accordingly.

## d.    Occurrence of Burn-Through

Both optical and acoustic signals and their frequency distributions when burn-through occurred are shown in Figure 22 and Figure 23.

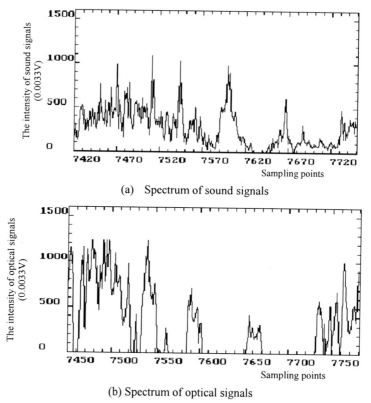

(a)    Spectrum of sound signals

(b) Spectrum of optical signals

Figure 22 Sound and optical signals when burn-through occurred.

Just before burn-through appeared, the intensity of both the optical and acoustic signals increased markedly. But the intensities of the signals in their spectrum mainly concentrated in a low frequency range. Then some holes appeared in the weld bead and the intensities of the signals declined abruptly.

Therefore, it could be seen that laser welding process stability could be monitored on line by comparing the intensities of the acoustic and optical signals. In order to diagnose the type of defect, the spectrum characteristics of the signals must be analysed.

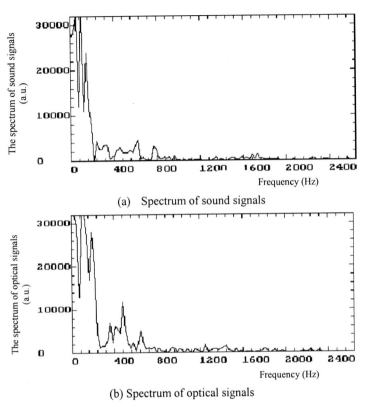

(a)   Spectrum of sound signals

(b) Spectrum of optical signals

Figure 23 Frequency distribution of sound (a) and optical (b) signals when burn-through occurred.

## 4.2   *Wavelet Analysis of Audible Acoustic Emission Signals*

Audible acoustic emission emerges as long as there is thermal motion of metal vapour even without plasma. On the other hand, in full penetration welds, sound pressure is enhanced due to the addition of acoustic emission from the plasma. This is a very useful characteristic. The quality of laser welding correlates closely with the behaviour of the melting pool and keyhole besides the plasma. When the melting pool and keyhole are damaged, the ejection of plasma from the keyhole is affected, resulting in the sharp drop in the intensity of acoustic emission. Analysis of the acoustic emission can infer characteristics of the melting pool and the keyhole.

### 4.2.1    Wavelet Decomposition of AE Signals

The AE signals for gap defect and misalignment defect are shown in Figure 24. The abscissa represents the index of signal sampling points while the ordinate is the intensity of the raw signal from the data acquisition board. It is a 12-bit A/D board with an input range of ±5V. The length of the signal is about 154,000 sampling points. The defects were located after comparing the test samples to the acquired signals. The first gap falls into the range of 56,000 - 70,000 sampling points while the second falls into the range of 96,000 - 109,000 sampling points.

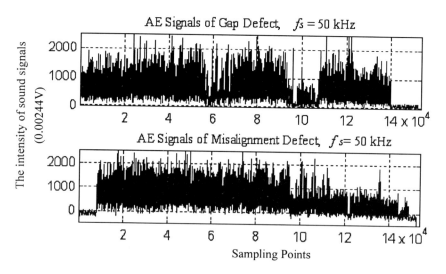

Figure 24 AE signals of laser welding with gap and Misalignment problem.

To make the analysis easier, the original signal was extracted from sampling points 53,000 to 63,000 which corresponded to the first gap defect. This was denoted as signal $s$. In signal $s$, the first gap is located from point 3300 to point 6100 and the second starts from point 9000. The signal $s$ was then decomposed to level 5 by the $db5$ (Daubechies, order 5) wavelet with $a5$ as the level 5 approximate signal, and $d5$, $d4$, $d3$ as level 5, level 4, level 3 detail signal respectively. The signal $s$ and its constituents are shown in Figure 25. According to the theory of wavelet decomposition, we have

$$s = a5 + d5 + d4 + d3 + d3 + d1 \tag{13}$$

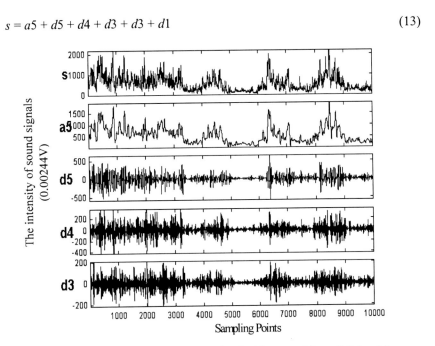

Figure 25 Wavelet decomposition of AE signal with gap problem, *db5*, level 5.

It should be noticed that *a5*, *d5*, *d4*, *d3*, *d2*, *d1* are reconstructed approximations and details. The goal of reconstruction is to help us analyse and understand signals intuitively. Down-sampling is applied to the decomposition signals ($ca_n$, $cd_n$). This simply means throwing away every odd index data point. In other words, the number of sampling points is halved. While approximations $ca_n$ and details $cd_n$ are reconstructed (we get $a_n$ and $d_n$), up-sampling is applied to make the length of $a_n$ and $d_n$ revert to the same length of signal *s*. If the length of sampling signal *Sig* is *L*, the length of $a_n$ and $d_n$ is also *L* while the length of $ca_n$ and $cd_n$ is reduced to $2^{-n}L$.

Based on the relation of frequency structure of wavelet decomposition, the frequency bandwidth of approximation and detail of level *l* are $\left[0, \frac{1}{2}f_s 2^{-l}\right]$ and $\left[\frac{1}{2}f_s 2^{-l}, \frac{1}{2}f_s 2^{-(l-1)}\right]$ respectively. It is noticed that the frequency band of every level is decomposed into two equal sub-bands, the detail and the approximation. The frequency bandwidths of $a6$, $d6$, $d5$, $d4$, $d3$ are [0, 390 Hz], [390 Hz, 781 Hz], [781 Hz, 1562 Hz], [1562 Hz, 3125 Hz], [3125 Hz, 6250 Hz] respectively. The result of wavelet translation is a series of decomposed signals belonging to different frequency bands.

### 4.2.2    Results of Wavelet Analysis of AE Signal

The plot in Figure 25 shows that $a5$ can be regarded as the result of passing signal $s$ through a low-pass filter. The intensity of $a5$ shows a noticeable drop at the gap defect while the intensity level maintains a minimum value of 450 during the normal welding state. This example indicates that there is potential for using a threshold technique to differentiate between acceptable and defective welds. The frequency bands of $d5$, $d4$ are [781 Hz, 1562 Hz], [1562 Hz, 3125 Hz] respectively. These belong to the lower frequency part of the signal. A noticeable drop in oscillation amplitude can be seen. In this example, the defect is only 20% of the amplitude of the normal weld. The amplitude of oscillation is a more robust indicator of defective welding than signal peak value alone as it is less susceptible to noise and interference.

These phenomena can be explained as follows. The full penetration weld was formed and AE signals were emitted steadily due to the effect of thermal vibration, plasma and keyhole behaviour. Therefore the sampling signal was expressed as a signal with definite intensity superimposed with high frequency oscillating components. When the gap appeared, the keyhole effect disappeared, as did the generation condition of plasma. All these weakened the AE signal. Visual inspection of the workpiece after welding indicated that the metal around the gap had melted and the behaviour of melting pool together with thermal vibration of workpiece were the main causes of AE signals. Compared with normal welds, the intensity of AE signals decreased.

Figure 26 shows the wavelet decomposition of AE signals, from normal welds and welds with a misalignment problem. The signals are extracted from original sampling signals. Each one has 10,000 sampling data from sections between [30,000 − 40,000] and [126,000 − 136,000] respectively.

The plots in Figure 26 (a) are for normal deep penetration welds, while those in Figure 26 (b) for welds with misalignment. Comparing the two sets of plots, it can be seen that the intensity of $a5$ for the defective weld drops sharply to 50% of that for normal welds. The same phenomenon can be found at *d*5 and *d*4.

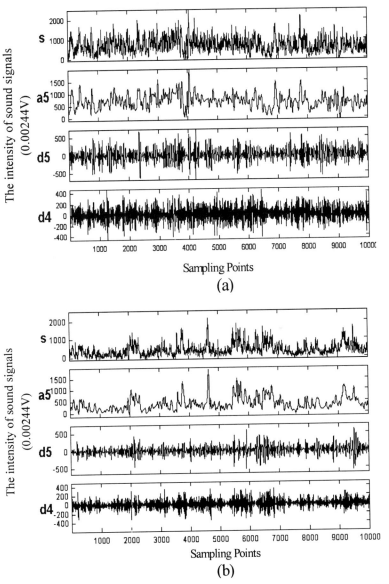

Figure 26 Wavelet decomposition of AE signal, compared with deep penetration welds (a) and welds with misalignment (b).

The existence of the fairly consistent difference in signal intensity in $a5$ and oscillation amplitude in $d5$ and $d4$ frequency bands suggested a very simple technique for discriminating full-penetration welds from welds with gap or misalignment defects.

## 4.3    *Definition and Applications of Detection Curve*

Based on the experimentation detailed above, the oscillation amplitude in $cd5$ frequency bands were treated as a symptom of defective laser welding. As the process of laser welding changed from normal to abnormal, the oscillation amplitude level dropped. After further signal processing, a detection curve to recognise the transformation of welding states was defined.

### 4.3.1    *Definition of Detection Curve*

A signal intensity moving average curve or *IMA* curve is defined, based on the wavelet decomposition component $cd_n$, as follows

$$IMA_{seg,shift}(i) = \frac{1}{seg} \sum_{m=1+i \times shift}^{seg+i \times shift} |cd_n(m)| \quad i = 0,1,2,3... \tag{14}$$

$m$    must    be    less    than    the    length    of    $cd_n$,    that    is $seg + i \times shift \leq length\,(cd_n)$. Hence the length of *IMA* curve is

$$Length(IMA) = round(\frac{length(cd_n) - seg}{shift}) + 1 \tag{15}$$

The corresponding relation between the point $i$ of *IMA* curve of $cd_n$ and the point $k$ of AE signal *Sig* is

$$IMA:\ i = round(\frac{2^{-n}k - seg}{shift}) + 1 \longleftrightarrow Sig: k \tag{16}$$

Laser welding is a high-speed welding process (welding speed faster than 2m/min). To ensure welding quality, the welding defect identified by a monitoring system should be limited to 1 mm. The response time of a monitoring system should be less than 30 ms if the welding speed is

2m/min. This is the minimum requirement for such a laser welding monitoring system. It should be noticed that the decomposition signal is down-sampled during wavelet transform, which means the data size of signal is halved. For example, the data size of *cd5* component will reduce to $2^{-5} = 1/32$ of original signal. A solution is to increase the sampling frequency $f_s$ to 50 kHz to meet the minimum number of sampling points for signal processing.

## 4.3.2   Example

According to the sampling frequency and processing capacity, *seg* was set to 32 and *shift* was set to 8. When a 1024-point data set (approximately 20*ms*) was sampled, the *db5* wavelet was applied to decompose the sampling signal to level 5 (The length of *cd5* component was $2^{-5} \times 1024 = 32$ samples). Then the *IMA* of the *cd5* component ([781 Hz, 1562 Hz]) was calculated. When the next 1024-point data set was sampled, it was decomposed to produce $\tilde{cd}5$ component. Then the $\tilde{cd}5$ component was appended to the old *cd5* component to make up a 64-point *cd5* component. The *IMA* was then calculated. This procedure was repeated for all sample signals in the entire welding process. The AE signals, *cd5* component and relevant *IMA* curves are shown in Figure 27 and Figure 28.

The plot in Figure 27 shows that the *IMA* curve of the *cd5* component maintains higher than 500 for a normal weld condition. There is a dramatic drop in the *IMA* curve as the gap emerges. The curve drops from 500 to 100. The drop is amplified and shown in Figure 28. In the plot, it can be seen that the *IMA* curve declines from sampling point 220.

Each gap is made up of two small holes and this is substantiated by the *IMA* curve. There is a jump between two minima of the *IMA* curve, which indicates that the welding state returns to normal after the gap was passed. However, the value of the *IMA* curve at this time is low compared with full penetration welds because the *IMA* curve is an average magnitude. It should be noticed that Gap2 is more distinct than Gap1 in AE signals because the width of Gap2 (0.12 mm) is larger than that of Gap1 (0.08 mm). This difference is also highlighted by the *IMA* curve too. The section of *IMA* curve with lower value and higher smoothness represents the case of the larger gap. The section of *IMA* curve with higher value and oscillation represents the case of the smaller gap. This implies that *IMA* curve can be used to indicate the size of the gap.

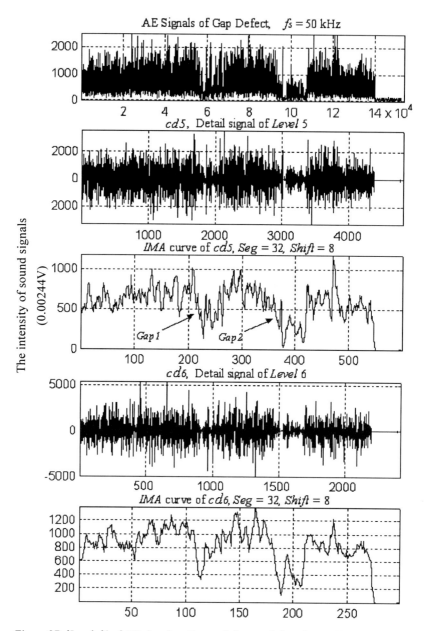

Figure 27 *d*5 and *d*6 of AE signals with gap defects and the relevant *IMA* curves (for original AE signals, the abscissa is the sampling point).

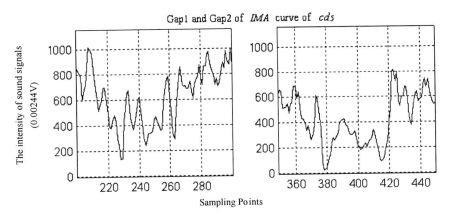

Figure 28 Zooming in of Gap1 and Gap2.

Likewise, the *IMA* curve of the *cd*6 component was calculated. As shown in Figure 27, the *IMA* curve of *cd*6 is much smoother than that of *cd*5. The general shapes of these two curves are similar although some details are not shown in the *IMA* curve of *cd*6.

The plot in Figure 29 shows the data from the experiments pertaining to the misalignment problem. The process is divided into three phases according to the amount of misalignment. At the beginning of the signals (*Phase I*), in normal welding state, the *IMA* curve maintains a minimum of 600 although the oscillation is very severe. In *phase II*, the amount of misalignment begins to increase as the welding process carries on. But the *IMA* curve still reaches 600 and keeps at the level with much more severe oscillation than *phase I*. The misalignment problem cannot be detected from the *IMA* curve at this time.

This shows that the change of AE signals is insignificant when the amount of misalignment is small. The difference of evaporation capacity between *phase I* and *phase II* is insignificant because the change of defocusing amount within ±0.5 mm along the optical axis still falls into the focal depth. At the same time, the change of workpiece surface appearance is not sufficient enough to change the shape of melt pool and keyhole. These are the reasons why misalignment problem cannot be detected from *IMA* curve in *phase II*.

As the welding process carries on (*phase III*), the amount of misalignment becomes bigger and bigger and causes the characteristics of melt pool and keyhole to change. It is noticed that there is a sharp drop at point A in the plot. It can be observed clearly that the *IMA* curve decreases almost linearly as the welding process carries on because the misalignment is a linear, gradual process.

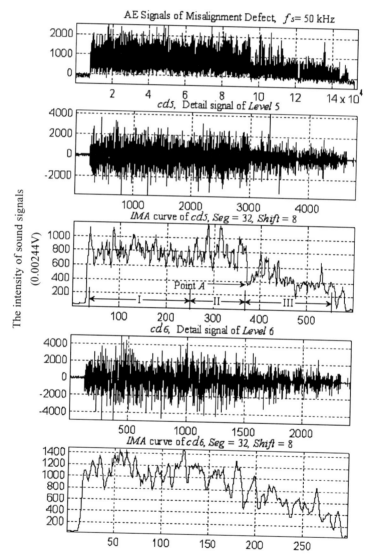

Figure 29 *d5* and *d6* of AE signals with misalignment defects and the relevant *IMA* curves (for original AE signals, the abscissa is the sampling points).

Simply comparing the *IMA* curve with a suitable threshold value, welds above the threshold value can be classified as normal full penetration and those below the threshold value are defective or partial.

In conclusion, the AE signal can play a role in detecting variables during process control because it has a high signal noise ratio (SNR), fast response and non-contact measurement characteristics. The generation of the AE signals can be attributed to thermal vibration induced by temperature gradient fields in the melt pool, sound pressure engendered by keyhole effect and plasma behaviour.

Signals can be decomposed into a series of constituents, which distribute over different frequency bands after wavelet transform. Further information processing can be carried out on selective components according to the given application. The intensity of low-frequency band (<781 Hz) and oscillation amplitude of frequency component ([781 Hz, 1562 Hz]) of the AE signals decrease dramatically when the welding defects occur. The same phenomena are observed in misalignment defect experiments too. The amplitude of oscillation is a more robust indicator of defective welding than signal peak value alone as it is less susceptible to noise and interference.

In this study, a detection curve (*IMA* curve) is defined, and can be used to recognise the transformation of welding states. The welds having gap defect or misalignment problem can be detected by measuring oscillation amplitude in the frequency band [781 Hz, 1562 Hz].

## 5. Real-Time Monitoring of Laser Welding by ANN

An artificial neural network (ANN) has the ability to partition the feature space into regions of complex shape and can be trained to recognise patterns that are separable in its feature space. Acoustic emission signals generated during different welding conditions are distinguishable in principle, and it is possible to detect weld faults using a neural network.

### 5.1 *Structure of the Neural Network*

#### 5.1.1 *Construction of Features*

Feature extraction is the foundation of pattern recognition. Signal processing is needed to extract features, which are the inputs of the neural network, from acoustic emission signals. For an efficient identification, the inputs of ANN should include as many features as possible.

Previous research shows that using only one component of wavelet analysis results of AE signals is insufficient to detect weld faults. The problem is that the threshold is non-adaptive and needs manual adjustment. More research work is needed to apply wavelet analysis results for automatic monitoring of laser welding. The welding information is distributed over the whole spectrum of AE signals. Better results could be achieved if every component of wavelet analysis results, i.e. a series of approximations and details, is taken into consideration. The main approach is described as follows.

*a.    Decomposing AE Signals with Wavelet Analysis, Using Wavelet Package Analysis if Necessary*

The AE signals are decomposed first to level $i$ with wavelet analysis, and then reconstructed to produce a reconstructed approximation $A_i$, and reconstructed details $D_1, D_2, \ldots, D_i$. The original signal $S$ can be presented as

$$S = A_i + D_i \ldots D_2 + D_1 \tag{17}$$

Based on the relation of frequency structure of wavelet decomposition, the frequency bandwidth of approximation and detail of level $l$ are $\left[0, \frac{1}{2} f_s 2^{-l}\right]$ and $\left[\frac{1}{2} f_s 2^{-l}, \frac{1}{2} f_s 2^{-(l-1)}\right]$ respectively, where $f_s$ is the sampling rate of signal $S$.

*b.    Computing Feature of Components such as IMA, RMS and Variance*

A feature vector $T'_{Eigen}$ can be written as follows,

$$T'_{Eigen} = [Eigen_{Ai}, Eigen_{Di}, \ldots, Eigen_{D2}, Eigen_{D1}]^T \tag{18}$$

*c.    Detecting Singularity Point of AE Signals*

The result is denoted as *Flag*. The value of *Flag* is 0 (without singularity) or 1(with singularity).

## d.   Combining All These Features to Form Input Feature Vector of ANN

The final feature vector $T_{Eigen}$ is

$$T_{Eigen} = [Eigen_{Ai}, Eigen_{Di}, ..., Eigen_{D2}, Eigen_{D1}, Flag]^T \tag{19}$$

Compared with the method of generating a detection curve from original signals, there is an advantage in forming an input feature vector. Since signals are decomposed into different frequency bands, the characteristics of these frequency bands can be easily detected.

### 5.1.2   Improvement of Features

The AE signal is processed every 1024 sampling points with wavelet analysis to construct a feature vector with 9 variables, including one approximation, seven details and one flag of singularity. The training set is constructed mainly from a typical gap fault experiment, partially from other experiments. The experimental data and target value are randomised such that the same pattern of data does not reappear in a sequence of recording. Target value 0.9 represents normal weld status while target value 0.1 represents fault weld status.

Training results show that the training time is longer than the prescribed epoch of 30000 with a very low convergence speed. Additionally, the validation result is not satisfactory. This situation is caused by insufficient input data to the ANN. The distribution of the AE signal may be scattered in its feature space. There are a considerable number of overlapped regions between data from different welding conditions. It is very difficult for the ANN to find the boundaries to efficiently separate the training data with insufficient data. The feature vector should be reconstructed to contain more information about the weld status.

Among all the eight approximation and details, the amplitudes of $D_1$ and $D_2$ are much smaller than other components. This indicates that the energies of these two components are very small and of negligible status.

The feature of each component is fluctuating during the welding process. The values of some component features are very small sometimes even when the welding status is quite good. It will be more reliable if historical data of the welding process can be integrated into the input vector of the ANN.

According to this concept, the feature vector is simplified and improved. $D_1$ and $D_2$ components are deleted from the feature vector to make the training process more efficient. Suppose *hist* points of past information will be added to feature vector, past information features are arranged at the end of current time features. The total length of the feature vector is then extended to $(hist+1) \times 7$. If there is no past information, (for example, at the beginning of welding process), the value is filled with 0.

The convergence speed of the ANN is improved markedly with the new feature vector, as shown in Table 4.

Table 4 The relationship between *hist* and training speed.

|            | *hist* = 0 | *hist* = 5 | *hist* = 10 | *hist* = 20 | *hist* = 20 |
|------------|-----------|-----------|------------|------------|------------|
| Epochs     | 30000     | 3182      | 1170       | 700        | 1091       |
| Error goal | 0.89%     | 0.5%      | 0.5%       | 0.5%       | 0.1%       |

## 5.1.3    BP Network Parameters and Effect of Different Features

The neural network used here is a backpropagation (BP) neural network with one hidden layer. The standard BP network is improved with adaptive learning rate, momentum and Nguyen-Widrow weight initialisation method. In most situations, there is no way to determine the best number of hidden nodes without training several networks and estimating the generalisation error of each. The number of hidden nodes is determined by experiments in this application. The parameters of BP neural network are listed as Table 5.

Table 5 BP network parameters.

| Learning rate | 0.01 | Learning rate variation factor | 0.7, 1.1 |
|---------------|------|-------------------------------|----------|
| Error goal | 0.1% | Max. epochs | 30000 |
| Trans func of hidden layer | logsig | Trans func of output layer | pureline |
| Momentum constant | 0.9 | Num. of hidden nodes | 100 |
| Num. of input nodes | $(hist+1) \times 7$ | Num. of output nodes | 1 |

The learning rate is adjusted in response to the change in the error. If the current error is more than the previous one, the learning rate is decreased by the factor of 0.7. If the current error is less than the previous one, the learning rate is increased by the factor of 1.1.

Two different methods of feature encoding are used. One is *IMA* and the other is *RMS*. The performances of two encoding methods are studied on the experimental data. The *correct, wrong* and *uncertain* are defined as follows:

- Correct: normal welding and the output of ANN is larger than 0.6; or abnormal welding and the output of ANN is less than 0.4.
- Wrong: normal welding and the output of ANN is less than 0.4; or abnormal welding and the output of ANN is larger than 0.6.
- Uncertain: the output of ANN is between 0.4 and 0.6.

The test results are outlined in Table 6 and Table 7. The performance of the two feature encoding techniques is almost the same although *IMA* looks a little better. Both results are acceptable with accuracy of 85%. But the non-accuracy parts (wrong and uncertain) are quite dissimilar with different neural network structures. In some cases, the inaccuracy and uncertainty is 3.6% and 8.9% respectively.

Table 6 Percentage of weld fault classification results using IMA.

| | | hist = 10 | hist = 20 | hist = 30 | hist = 40 |
|---|---|---|---|---|---|
| Neuron Num = 100 | Accuracy | 87.0% | 87.5% | 87.2% | 84.5% |
| | Error rate | 5.0% | 3.6% | 4.3% | 4.6% |
| | Uncertainty | 8.0% | 8.9% | 8.5% | 10.9% |
| Neuron Num = 150 | Accuracy | 85.7% | 84.2% | 85.8% | 81.8% |
| | Error rate | 4.3% | 4.4% | 5.2% | 4.5% |
| | Uncertainty | 10.0% | 11.4% | 9.0% | 13.7% |
| Neuron Num = 200 | Accuracy | 84.2% | 84.2% | 83.8% | 83.3% |
| | Error rate | 4.9% | 4.8% | 6.7% | 4.0% |
| | Uncertainty | 10.9% | 11.0% | 9.5% | 12.7% |
| Neuron Num = 250 | Accuracy | 84.0% | 82.8% | 85.4% | 82.0% |
| | Error rate | 5.0% | 5.0% | 5.2% | 6.6% |
| | Uncertainty | 11.0% | 12.2% | 9.4% | 11.4% |

Table 7 Percentage of weld fault classification results using RMS.

|  |  | *hist* = 10 | *hist* = 20 | *hist* = 30 | *hist* = 40 |
|---|---|---|---|---|---|
| Neuron Num = 100 | Accuracy | 86.8% | 86.3% | 85.7% | 84.5% |
|  | Error rate | 7.0% | 5.3% | 4.3% | 4.5% |
|  | Uncertainty | 6.2% | 8.4% | 10.0% | 11.0% |
| Neuron Num = 150 | Accuracy | 86.6% | 82.4% | 88.6% | 84.2% |
|  | Error rate | 6.7% | 5.5% | 4.8% | 5.0% |
|  | Uncertainty | 6.7% | 12.1% | 6.6% | 10.8% |
| Neuron Num = 200 | Accuracy | 84.6% | 82.8% | 84.5% | 83.7% |
|  | Error rate | 7.7% | 5.8% | 5.0% | 3.6% |
|  | Uncertainty | 7.7% | 11.4% | 10.5% | 12.7% |
| Neuron Num = 250 | Accuracy | 83.2% | 81.5% | 86.6% | 81.9% |
|  | Error rate | 6.3% | 6.4% | 5.0% | 4.1% |
|  | Uncertainty | 10.5% | 12.1% | 8.4% | 14.0% |

When the output of ANN is between 0.4 and 0.6, ANN fails to identify the welding status. It is difficult to say if it is normal or abnormal. Practically, this output could be treated as a warning signal. The training and validation results are shown in Figure 30.

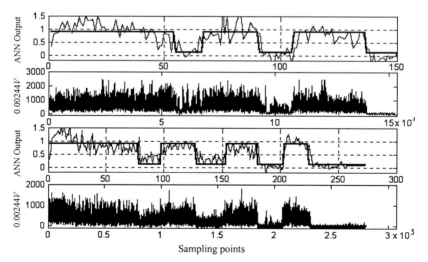

Figure 30 Training and validation results of ANN (the bold line is target, and the normal line is output of ANN).

## 5.2 *Performance of the Neural Network*

Here, the focus is on generalisation of the trained network with data from other experiments. The sole aim is to establish whether or not the network has any potential for broader applicability.

The first part of validation is concerned with misalignment weld fault experiments. Two sets of experimental data are fed into the trained network, as shown in Figure 31. Since the historical information is treated as part of the feature, the initial output of network is unreliable because the feature vector is filled with zero at the beginning when there is no historical information. The validation results of the network are generally satisfying for the misalignment fault.

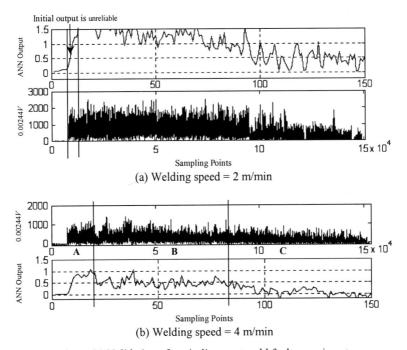

Figure 31 Validation of a misalignment weld fault experiment.

The second part of investigation involves presenting the trained network with data from gap weld fault experiments, as shown in Figure 32. Most of output values of the network are uncertain while there is gap fault. The reason is that the gap is very small in this experiment. The neural network does not generate satisfactory results for small gap weld fault.

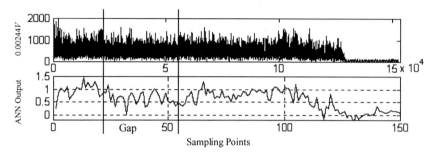

Figure 32 Validation of a gap weld fault experiment.

Data from normal weld experiments are also fed into the trained network to validate its ability, as shown in Figure 33.

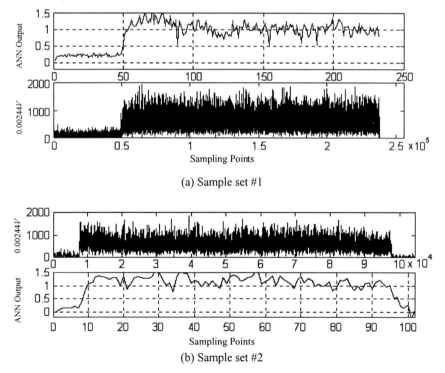

(a) Sample set #1

(b) Sample set #2

Figure 33 Validation of a normal weld experiment.

The results are satisfactory. The accuracy is nearly 100% although there are outputs of uncertainty in several network judgements.

## 5.3 *Discussion*

Based on the above results, the following conclusions can be drawn on the application of BP ANN to laser welding fault identification.

- The initial output of ANN is unreliable since historical information is included in the feature vector. The convincing identification can be obtained after *hist* information is fed into the neural network.
- The threshold of normal and abnormal is 0.6 and 0.4 respectively. The output between 0.4 and 0.6 can be treated as a warning signal since the ANN cannot identify the welding status.
- High accuracy can be achieved when the ANN is applied to normal welding. So a very low possibility of misjudgement on normal welding is guaranteed.
- A trained network can detect the welding fault due to the gap in the butt joint. However, when the gap is small, the network cannot maintain a stable output below 0.4, and the output is fluctuating around 0.5.
- Misalignment welding faults can be detected by a trained network. Generally, the output of the network can be divided into three phases. The output that is fluctuating around 0.5 in phase II indicates that the welding status is also undulating under a certain amount of misalignment. The network can maintain a stable output below 0.4 only if the amount of misalignment is bigger than a certain value.

## 6. Conclusions

In this study, a defect detecting system for laser welding, based on the optical and acoustic emissions from the laser-induced plasma, has been implemented as a workable prototype. The following conclusions can be made:

1. The existence of laser-induced plasma during laser welding of thin steel plate is proved.
2. The correlation between the optical and acoustic emissions from the plasma and laser welding quality is obtained.

3. FFT, wavelet analysis and *IMA* curves are used to analyse the signals when various defects occur in laser welding.
4. A neural network can be designed to reliably identify laser welding defects.

# References

1. Weber, H. "High average power solid state lasers for materials processing and fibre transmission". Laser Applications for Mechanical Industry (Editors: Martellucci, S. et al.), Kluwer Academic Publishers, 1993, pp. 11-29.

2. Ehlers, B., Walz, U., et al. "Cylindrical zoom optics for high power diode lasers", Proceedings of SPIE - The International Society for Optical Engineering. Vol. 3929, 2000, USA, pp. 185-192.

3. Ehlers, B., Herfurth, H.J., et al. "Hardening and welding with high power diode lasers", Proceedings of SPIE - The International Society for Optical Engineering, Vol. 3945, 2000, pp. 63-70.

4. Azuma, K., and Ikemoto, K. "Laser welding technology for joining different sheet metals for one piece stamping". Laser Applications for Mechanical Industry (Editors: Martellucci, S. et al.), Kluwer Academic Publishers, 1993, pp. 219-233.

5. Steen, W. M. "Laser process automation", Laser Applications for Mechanical Industry (Editors: Martellucci, S. et al.), Kluwer Academic Publishers, 1993, pp. 31-43.

6. Willmott, N.F., Hinnerd, R., and Steen, W.M. "Keyhole/plasma sensing system for laser welding control system", Proceedings of ICALEO'88, 1988, pp. 109-118.

7. Steen, W.M. "Monitoring of laser material processes", SPIE High Power Lasers and Their Industrial Applications, Proceedings of SPIE, V668, 1986, pp. 160-166

8. Steen, W.M. "In process beam monitoring", SPIE Laser Processing: Fundamentals, Applications, and Systems Engineering, Proc. SPIE, V668, 1986, pp. 37-44.

9. Weeter, L., and Albright, C. "The effect of full penetration on laser-induced stress wave emissions during laser spot welding", Mater. Eval. 45, 1987, pp. 353-357.

10. Hamann, C., Rosen, H.G., and LaBiger, B. "Acoustic emission and its application to laser spot welding", SPIE High Powers Lasers and Laser Machining Technology, Proc. SPIE, Vol. 1132, 1989, pp. 275-281.

11. Ton, M.C. "Noncontact acoustic emission monitoring of laser beam welding", Welding Journal, Vol. 64(9), 1985, pp. 43-48.

12. Li, L., and Steen, W.M. "Non-contact acoustic emission monitoring during laser processing", ICALEO'92, Vol. 75, pp. 719-728.

13. Mombo-Caristan, J.C., Koch, M., and Prange, W. "Seam geometry monitoring for tailored welded blanks", ICALEO'91, Vol. 74, pp. 123-132.

14. Gu, H., and Duley, W.W. "Statistical approach to acoustic monitoring of laser welding", J. Phys. D 29, 1996, pp. 556-560.

15. Gu, H., and Duley, W.W. "Analysis of acoustic signals detected from different locations during laser beam welding of steel sheet", ICALEO'96, pp. B40-B48.

16. Gu, H. and Duley, W.W. "Resonant acoustic emission during laser welding of metals", J. Phys. D, 29, 1996, pp. 550-555.

17. Farson, D., Hillsley, K., et al. "Frequency-time characteristics of air-borne signals from laser welds", J. Laser Appl. 8, 1996, pp. 33-42.

18. Nava-Rudiger, E., and Houlot, M. "Integration of real time quality control systems in a welding process", J. Laser Appl. 9,1997, pp. 95-102

19. Lankalapalli, K.N., Tu, J.F., et al. "Laser weld penetration estimation using temperature measurements", Journal of Manufacturing Science and Engineering, 121, 1999, pp. 179-188.

20. Rockstroh, T., and Mazuder, J. "Spectroscopic studies of plasma during cw laser materials interaction", J. Appl. Phys. 61, 1983, pp. 917-923.

21. Muller, R. "Real time monitoring of laser weld plum temperature and species concentration", ICALEO'96, pp. B68-B75.

22. Beyer, E., Maischner, D., et al "A neural network to analyze plasma fluctuations with the aim to determine the degree of full penetration in laser welding", ICALEO'94, Vol. 79, pp. 51-57.

23. Kluft, W., Boerger, P., et al. "On-line monitoring of laser welding of sheet metal by special evaluation of plasma radiation", Prometec GmbH, Aachen, Germany, 1996.

24. Kaierle, S., Dahmen, M., et al. "Autonomous manufacturing: planning and control in laser beam welding", ICALEO'96, pp. 154-163.

25. Li, L., Qi, N., et al. "On-line laser welding sensing for quality control", ICALEO'90, pp. 411-421.

26. Li, L. "Sensor development for in-process quality inspection and optimization of high speed laser can welding process", LAMP'92, pp. 421-426.

27. Sun, A., and Kannatey-Asibu, E. "Sensor systems for real-time monitoring of laser weld quality", Journal of Laser Applications, Vol. 11(4), pp. 153-167.

28. Chen, H., and Li, L. "Laser process monitoring with dual wavelength optical sensors", ICALEO'91, pp. 113-122.

29. Haran, F., Hand, D., et al. "Process control in laser welding utilizing optical signal oscillations", ICALEO'96, pp. B49-B57.

30. Jurca, M., Paper 93 LA 030, Jurca Optoelektronik GmbH, Germany, 1993.

31. Hanting, J. and Aiqing, D. "Real time measurement method of laser welding process with acoustic signals and plasma radiation", LAMP'92, pp. 457-459.

32. Dixon, R., and Lewis, G. "Electron emission and plasma formation during laser beam welding", Welding Journal, 64(3), 1985, pp. 71s-78s.

33. Gatzweiler, W., Maischner, D., et al. "On-line plasma diagnostics for process control in welding with $CO_2$ lasers", SPIE High Power $CO_2$ Laser Systems and Applications, Proceedings of SPIE Vol. 1020, 1988, pp. 142-148.

34. Zeng, H., Zhou, Z.D., Chen, Y.P., Luo, H., and Hu, L.J. "Wavelet analysis of acoustic emission signals and quality control in laser welding", Journal of Laser Applications, 2001, 13(4), pp. 167-173.

# INDEX